# The Legend of **GRAINGER**®

The Legend of
GRAINGER®

Jeffrey L. Rodengen

Edited by Melody Maysonet
Design and layout by Sandy Cruz

Publisher's Cataloging-in-Publication

Rodengen, Jeffrey L.
　The legend of Grainger/Jeffrey L. Rodengen.
– 1st ed.
　p. cm.
　Includes bibliographical references (p.) and index.
　　LCCN 97-062152
　　ISBN 0-945903-43-X

　1. W. W. Grainger, Inc. 2. Machinery industry – United States – History. 3. Industrial equipment – Maintenance and repair – Equipment and supplies. 4. Machinery – Equipment and supplies.
I. Title.

HD9705.U64G73 2002　　　338.7'6218
　　　　　　　　　　　　　　QBI21-76

Write Stuff Enterprises, Inc.
1001 South Andrews Avenue
Second Floor
Fort Lauderdale, FL 33316
1-800-900-Book (1-800-900-2665)
954-462-6657
www.writestuffbooks.com

Copyright © 2002 by Write Stuff Enterprises, Inc. All rights reserved. No part of this book may be reproduced or transmitted in any form by any means, electronic or mechanical, including photocopying and recording, or by any information storage or retrieval system, without permission in writing from the publisher.

Completely produced in the
United States of America
10 9 8 7 6 5 4 3 2 1

TABLE OF CONTENTS

Introduction iv
Acknowledgments viii

Chapter I	A Motor-vated Man: William W. Grainger (1927)	10
Chapter II	Fast Rise and the Big Crash (1927–1929)	14
Chapter III	Expansion during Depression (1930–1939)	18
Chapter IV	The Grainger Homefront (1940–1945)	30
Chapter V	Scarcity and Boom (1946–1949)	40
Chapter VI	Staying Ahead (1950–1959)	48
Chapter VII	Grainger Goes Public (1960–1969)	58
Chapter VIII	Expansion and Realignment (1970–1979)	70
Chapter IX	Market Penetration (1980–1989)	80
Chapter X	Strategy for Success (1990–1995)	96
Chapter XI	A Competitive Advantage (1996–2002)	114
Appendix A	Officers of W. W. Grainger, Inc.	144
Appendix B	Executive Council as of May 2002	146
Appendix C	Officers as of May 2002	147
Appendix D	Directors of W. W. Grainger, Inc.	148
Appendix E	Steady Sales Growth through the Years	150

Notes to Sources 152
Index 166

INTRODUCTION

IT TAKES MORE THAN AN ENTREPRENEURIAL spirit, hard work, and a favorable market for a company to attain 75 years of success, and W. W. Grainger, Inc., has discovered the right formula. The remarkable story of how Grainger has managed to survive and even thrive through Depression, recessions, wars, and an ever changing market begins in 1927, when a young electrical engineer named William Wallace Grainger opened a business in Chicago distributing wholesale electric motors and other industrial items. The first eight-page catalog mailed to prospects and customers was called the *MotorBook*. The catalog would grow to 4,000 pages, listing 100,000 items.

Bill Grainger started his business just after Americans had begun using electric motors to power everything from milking machines and fans to pumps and compressors. Even during the Great Depression, the products he delivered were in high demand, and the fact that he devoted himself to giving customers quality products and service at low prices kept his business alive when so many others were surrendering to the devastating economic conditions.

Despite a down year in 1932, the company experienced sales growth during the Depression along with its reputation for fast delivery and honest service. In 1933 it opened its first branch, in Philadelphia, and in 1938 it began manufacturing fans and blowers. It also began offering its own line of Dayton-brand motors and allied products. At a time when most Americans were grateful to have any kind of job, Grainger protected its employees, choosing to shorten work hours rather than discharge anyone. By the end of 1939, it had opened 17 branches across the country and had begun employing outside salesmen for each branch.

Grainger had always offered generous benefits to employees, but the Profit Sharing Plan it inaugurated in 1941 was truly extraordinary. Grainger contributed part of its net profits each year into a profit sharing trust fund that was then allocated to eligible employees' accounts and paid out at their retirement. In 1941, fewer than 400 companies in the entire United States offered such a plan.

Rationing during World War II caused the company to diversify into selling "non-priority" items such as dinnerware, pictures, cookbooks, toys, games, playing cards, and table lamps. The restrictions were lifted in 1945, and Grainger flourished in the postwar boom years as consumers rushed to buy the motors, fans, and household items they had been forced to do without. These were years of unparalleled prosperity for the United States—and for Grainger—and the company was quick to reward employees for a job well done—through the Profit Sharing Plan and numerous other incentives. By 1959, Grainger's sales had increased to $33.3 million, from $27.7 million the year before, and it had 64 branches and more than 600 employees.

From its founding, W. W. Grainger, Inc., had been a family company. Bill's sister Margaret had been the first company employee, and in 1952, Bill's son, David, an engineer like his father, began working for Grainger full time. Bill Grainger extended the family ambience to the entire workforce. Every Christmas, he sent employees a signed Christmas card and a ham, and he took the time to get to know as many of them as he could.

In 1952, Bill and his wife, Hally, established The Grainger Charitable Trust. It was later renamed The Grainger Foundation and has been managed and supported by David and his wife, Juli. Through the years, more than $125 million has been provided in support of educational, health care, religious, museum, and orchestral programs, as well as a myriad of other worthwhile local and national projects.

Employees were respected and well cared for and never had cause to question the company's integrity, for Bill Grainger established an unswerving code of ethics that he passed on to his colleagues, employees, suppliers, and to his son, David, who would succeed his father as Grainger's leader and pass on that same code of ethics to his colleagues, employees, suppliers, and eventual successors.

The company grew steadily during the 1960s, achieving record sales every year with an ever expanding line of products. By 1967, Grainger had 92 branches and nearly 1,200 employees, and its sales exceeded $89 million.

That was the year Grainger became a public company. Bill Grainger retired from active management in January 1968, passing the leadership reins to David Grainger as chairman of the board and to Edward Schmidt as president. Bill's retirement and the company's new presence as a public company signified the end of an era, but the founding principles remained intact. Grainger was and always would be a company that provided customers with the best-quality products and highest standard of service, that treated employees and suppliers with the utmost respect, and that gave back to the communities it served.

The 1970s was a difficult decade for the American economy. Many businesses remained stagnant or suffered losses, but Grainger continued to expand, diversify, and restructure. Its local branch network and territory sales representatives enabled it to serve broader markets while still maintaining individualized service. Through an innovative supply chain structure and technologically advanced inventory management and processing at its branches and distribution centers, the company was able to maintain its reputation for efficient service.

Grainger celebrated its 50th anniversary in 1977. The company had 141 branches, 7,600 products featured in its catalog, 570,000 customers, and sales of $498.5 million (a 25 percent increase over the previous year). By then, Ed Schmidt had retired and the presidential reins passed to David Grainger. At the 50th anniversary celebration, Bill and David Grainger received a standing ovation for their contributions to the lives of employees, customers, suppliers, and the electrical industry as a whole. Five years later, in October 1982, William Wallace Grainger passed away at age 87. The following year, Wiley Caldwell became president, and David Grainger was chairman and CEO.

Throughout the 1980s, Grainger remained a model of business success. Its new technologically advanced automated storage and retrieval system at the Central Distribution Center in Chicago increased productivity and accuracy, and a computerized branch order processing and inventory control system at all the branches enhanced customer service and branch productivity, improved inventory control, and made better use of warehouse space. In 1983, the company had grown enough to open a 1.4-million-square-foot distribution center in Kansas City. The following year, the $1 billion sales milestone was reached.

In 1986 Grainger's long-term planning strategy led it to sell its Manufacturing Group to better concentrate on its core competency of distribution—helping businesses minimize expenses by offering one-stop shopping for maintenance, repair, and operating (MRO) supplies. A marketing study in 1986 revealed that Grainger held only 2 percent of the MRO market, and the company set about greatly expanding its network of 200 branches by saturating metropolitan markets with easy-access branches. By the end of the decade, Grainger had branches in all 50 states, 311 in all.

A number of initiatives in the 1990s helped Grainger grow sales in an increasingly competitive, fast-paced environment while maintaining its high standards and improving customer satisfaction. The company started a special selling effort directed at large corporations. Those companies learned that they could lower MRO procurement costs by outsourcing to Grainger. The company also increased its marketing efforts and made acquisitions to fill out its product offerings in such areas as sanitary, safety, and janitorial supplies. It also opened branches in Puerto Rico and Mexico and acquired a 178-branch industrial supply distributor in Canada called Acklands Limited, which became Acklands - Grainger Inc. Ongoing training and state-of-the-art computer systems and technology helped Grainger stay on the cutting edge of customer service and efficiency. Also, by expanding its network of distribution centers, Grainger was better able to respond to customers' diverse needs.

The company transitioned to new leadership in the 1990s. Wiley Caldwell retired in 1992, making way for Richard Keyser to become president and COO in 1994, CEO in 1995, and chairman of the board in 1997. David Grainger continued as senior chairman of the board. Meanwhile, Grainger was becoming more and more agile in the increasingly competitive marketplace by reexamining itself and restructuring and by embracing burgeoning technologies while continuing to expand its bricks-and-mortar network. Grainger.com, for instance, which debuted in 1995, was way ahead of the competition and quickly became a model of success in business-to-business Internet commerce. The company also created new business units and restructured operations to better focus on domestic branch operations (Grainger Industrial Supply), replacement parts (Grainger Parts), safety and MRO supplies (Lab Safety Supply), consulting (Grainger Consulting Services), large companies' procurement needs (Grainger Integrated Supply), and procurement needs for large companies that needed materials management services (Grainger Custom Solutions). All of Grainger's business units were designed to give businesses and institutions what they wanted, when they needed it, at a lower cost than if they went through other distribution channels.

As it had since its founding, Grainger evoked loyalty among employees by providing a safe, pleasant workplace, incentives, benefits, and professional training. Employees embraced the company's values—agility, empowerment and accountability, ethics and integrity, having fun, learning, and teamwork—and Grainger's suppliers benefited mutually by partnering with a company that so deeply prized ethics and integrity.

By the turn of the century, Grainger had secured its position as the leading North American source of broad-line MRO supplies. Richard Keyser, chairman and CEO, along with Wesley Clark, who had become president and COO in 2001, were committed to upholding the cultural and business traditions that had made W. W. Grainger, Inc., such an enduring success: creating innovative processes, embracing new technologies, keeping a strong balance sheet, honing the skills of its talented employees, and constantly reexamining itself.

The Legend of Grainger stands as a testament to the integrity and industrious nature of William W. Grainger, as well as that of the people with whom he chose to work and those who have succeeded him.

ACKNOWLEDGMENTS

A GREAT NUMBER OF PEOple assisted in the research, preparation, and publication of *The Legend of Grainger*.

The principal research and narrative time line were the work of our research assistant Richard Buskin, whose efforts went a long way toward making this book a success.

This book would not have been possible without the vivid and forthright recollections of David Grainger, senior chairman of the board and retired president and CEO. Micheal Murray, retired vice president of administrative services, was instrumental in the project's development, and Philip Lippert, vice president of administrative services, lent his tireless support to see the book's completion. Mary Haider, records manager, was extremely helpful in collecting archival photos, as were Renee Young, employee publication consultant; Mary Weil, executive assistant; and Karen Sahm, employee communications writer and photographer.

Many other W. W. Grainger, Inc., executives, employees, retirees, and suppliers greatly enriched the book by discussing their experiences at or with the company. The author extends particular gratitude to these men and women for their candid recollections and anecdotes: Rick Adams, vice president of supply chain development; Jim Baisley, retired senior vice president and general counsel; Edward Bender, retired manager of information technology; Donald Bielinski, retired group president; Wiley Caldwell, retired president; Y. C. Chen, vice president of supply chain services; Barbara Chilson, former president of Grainger Parts; Wesley Clark, president and chief

operating officer; Robert Collins, retired vice president of branch operations; Douglas Cumming, retired chairman of Acklands - Grainger Inc.; Patrick Davidson, vice president for branch services; Timothy Ferrarell, senior vice president for enterprise systems; Lee Flory, retired vice president and secretary; Jere Fluno, retired vice chairman; Bob Gideon, employee programs manager; Gary Goberville, former vice president of human resources; Rich Greenlee, regional operations manager; Daniel Hamburger, former president of Orderzone.com; Douglas Harrison, president of Acklands - Grainger Inc.; Richard Heiman, president of Campbell, a division of Scott Fetzer; Nancy Hobor, vice president for communications and investor relations; John Howard, senior vice president and general counsel; Irene Imus, retired payment process service specialist; Dennis Jensen, retired vice president of field operations; Richard Keyser, chairman and CEO; Michael Kight, vice president of Grainger Integrated Supply; Ken Kirsner, corporate secretary; Jim Lindemann, CEO of Emerson Motor Company; Fred Loepp, vice president for production management; Larry Loizzo, vice president, and president of Lab Safety Supply; P. Ogden Loux, senior vice president of finance and CFO; Mark MacDonald, business unit manager of Brady Corporation; Tracy MacMillan, vice president of communications for North Safety; Connie Macias, retired data entry employee; Joseph Malak, chairman and retired president of Cleveland Sales; Thomas Malak, president of Cleveland Sales; Max Mielecki, retired vice president of advertising;

Robert Pappano, retired vice president of financial reporting; Earl Pope, retired manager of the tabulating department; Richard Quast, retired vice president for real estate; Joe Ramos, group president of Rubbermaid; Benedetto Randazzo, vice president of Latin American operations; George Rimnac, vice president and chief technologist; James Ryan, executive vice president of marketing, sales, and service; Angie Salazar, data entry and credit services manager; Edward F. Schmidt, retired president (courtesy of audio tape from Newstrack, Inc., Executive Tape Service); John Schweig, senior vice president for business development and international; Bob Slanicky, retired director of risk management; James Slavik, board member; John Slayton, senior vice president for supply chain management; Michael Tellor, president of Rust-Oleum; James Tenzillo, director of product services; Andrew Thomas, director of strategy; Robert Thrush, vice president of sales; Nancy Thurber, director of benefits; Peter Torrenti, former president of Grainger Integrated Supply; Paul Wallace, retired vice president for financial services; Phil West, vice president and treasurer; Robert Wiggins, retired vice president of sales; Dale Woods, retired consulting manager and former sales manager; and Fritz Zeck, president of Cooper Lighting Company.

As always, special thanks are extended to the dedicated staff at Write Stuff Enterprises, Inc.: Richard F. Hubbard, executive author; Jon VanZile, executive editor; Melody Maysonet, senior editor; Heather Deeley, assistant editor; Bonnie Freeman, copyeditor; Mary Aaron, transcriptionist; Barbara Koch, indexer; Sandy Cruz, senior art director; Rachelle Donley, Wendy Iverson, and Dennis Shockley, art directors; Bruce Borich, production manager; Marianne Roberts, vice president of administration; Sherry Hasso, bookkeeper; Linda Edell, executive assistant to the author; Lars Jessen, director of worldwide marketing; Joel Colby, sales and promotions manager; Rory Schmer, distribution supervisor; and Jennifer Walter, administrative assistant.

A visionary, William W. Grainger founded his wholesale electrical motor distribution business in 1927 at age 32. This portrait was painted circa 1960.

CHAPTER ONE

A MOTOR-VATED MAN: WILLIAM W. GRAINGER

1927: THE FOUNDING OF GRAINGER

WILLIAM WALLACE GRAINGER was the ultimate ideas man. He had ideas about everything in his line of business—about what items to stock and how to market them; how to evolve, expand, diversify, and adapt to new market conditions; how to capitalize on the boom times and prosper during periods of social and economic hardship; and how to earn the respect and loyalty of an ever expanding workforce.

"He always had new ideas about how to sell more motors," recalled Grainger's son, David, who, like his father, was an engineer. "Motors, motors, motors. He lived to sell a lot of motors."[1]

Of the thousands of ideas that ran through Bill Grainger's head, his most influential conception was to set up a wholesale electrical motor sales and distribution business. Seventy-five years later and following an evolutionary process that has wrought radical change, the company that bears his name still adheres to Grainger's basic principles: provide customers with the best-quality products and highest standard of service, take care of the workforce, and give something back to the community.

"He liked people," David Grainger said of his father. "He was relaxed about them. He hired people whom he trusted, and, while his own role was that of the benevolent entrepreneur, he felt complete security placing responsibility in their hands."[2]

"He didn't have any airs about being the owner of the company," added Dick Quast, a Grainger vice president. "He was a down-to-earth guy, and everybody liked him."[3]

Learning the Ropes

Born in Chicago on July 5, 1895, William W. Grainger displayed an entrepreneurial flair from an early age. When he was in second grade, the family purchased a 40-acre farm near the resort town of Volo, Illinois, and shortly thereafter, when not attending the one-room school that was located a one-mile walk from his home, young Bill was turning his first profit by digging up earthworms and selling them to local fishermen.

Bill graduated from Crane Junior College in 1915, and when World War I broke out, he joined the U.S. Navy and installed radio communication equipment on Navy ships. "That was back when radios used vacuum tubes and often were unreliable," said David Grainger.[4]

During his two years in the service, Bill achieved the highest enlisted rank of chief petty

---

W. W. Grainger, Inc., originally operated out of a single downtown Chicago office. By the end of the century, the company would have a significant presence throughout much of North America.

Near right: Hally Ward was the bookkeeper at Wagner Electric when she met Bill Grainger. They were married in 1923.

Far right: Margaret Grainger, sister of Bill Grainger, was W. W. Grainger, Inc.'s first employee. She played various roles in the company, including branch manager at Philadelphia (the first branch outside Chicago) and company secretary, until her retirement in 1953.

Below: At first, W. W. Grainger, Inc., generated sales through postcard mailers and the *MotorBook* catalog. The first issue of *MotorBook* was more like a pamphlet than a catalog, containing a mere 41 items within its eight pages.

officer, and when the war ended in 1918, he was asked to stay on as a lieutenant commander. He declined the offer, instead choosing to complete his education. In 1919 he emerged with a bachelor of science degree in electrical engineering from the University of Illinois at Champaign-Urbana.

Among Grainger's first jobs was designing electric motors for such manufacturers as General Electric, Master, Sunlight, Peerless, Brown-Brochmeyer, and Wagner. This work was uninspiring, however, and neither was he fulfilled by a highly successful stint as a motor salesman.

Along the way, he was gaining invaluable experience relating to the motor business. His work at Wagner Electric was especially notable in one respect. It was there that Bill met the woman who would be his wife for 59 years. Her name was Hally Ward, and she was Wagner's cashier and director of female personnel.

"My dad always asked a lot of questions," said David Grainger, recalling stories of his parents' early encounters. "He was very curious. People would say, 'Won't you leave me alone? If you want the answers, ask Miss Ward.' She had all the answers and she was outstandingly smart."[5]

Bill and Hally married in 1923 and would remain together until Bill's passing on October 9, 1982, just four weeks before Hally died. David said his mother did all of the bookkeeping for the family right up to three or four years before her death and her methods were impeccable:

She had this black book, and that black book was all of their investments, all of everything. She once got the bank to admit that they'd made an error on one of their statements. The error was a penny. So she was famous with [Grainger's] accounting firm of Alexander Grant & Company, and the word was out: "Don't argue with her." Junior accountants working with my mother would be told, "If what she says is different from what you've got, you're wrong, and you'll have to figure out why." That's quite a reputation.[6]

By the mid-1920s Bill was considering ways to branch out on his own. He observed the industrial application of electric motors, from master motors powering an entire shop to smaller electric motors powering individual machines. He also became increasingly aware of the difficulty electric motor shops had in obtaining motors quickly and in small quantities. Already having realized a need for an efficient wholesale business, Bill resigned when his employer balked at the large sales commissions he was earning.

This wheel arbor adaptor appeared on Grainger's first postcard mailer as part of a kit of electric motor accessories. Then and now manufactured by Clesco Manufacturing, it is one of thousands of products featured in Grainger's 2002 catalog.

"He was making more money than the sales manager," said David Grainger. "So the sales manager said, 'That's no good. I'm going to cut you back.' My dad said, 'That's fine because I'm out of here!' I guess that's when he decided to start the company."[7]

Young Entrepreneur

It was 1927, the year of David Grainger's birth. At the age of 32, Bill had become a father for the second time. (His oldest child, Barbara, was born in 1923.) He secured office space within a loft on 22nd Street (now known as Cermak Road), just south of downtown Chicago, and in September Bill and his sister Margaret opened the first office of W. W. Grainger, Inc. As the only employees of what was principally a mail-order business, they would receive and fill orders generated by postcard mailers and an eight-page, 41-item catalog called *MotorBook*.

The electric industry was evolving rapidly, yet the booming economy would soon be brought to its knees by the market crash of October 1929. Nonetheless, Bill Grainger would persevere during the most challenging years in the history of American business.

Bill Grainger (left) and his school chum Mark Graham, who ran Grainger's manufacturing department for a time, circa 1935

CHAPTER TWO

FAST RISE AND THE BIG CRASH

1927–1929

CHICAGO DURING THE LATE 1920s was hardly a city for the fainthearted. It was the era of Prohibition, speakeasies, bloody gangster warfare, and the Saint Valentine's Day Massacre. Chicago's mayor, Big Bill Thompson, left the running of the city largely to crime overlord Al Capone, resulting in a four-year spree of 215 gang killings without a single conviction. Amazingly, none of this bloodshed deterred people from flocking to the nation's second-largest metropolis.[1]

Throughout the 1920s, Chicago experienced immense building activity to accommodate a population that would rise to 3.5 million by the decade's end. In the midst of such downtown landmarks as the Wrigley Building, Tribune Tower, and Michigan Avenue bridge, skyscrapers began to rise. The value of new construction exploded to a staggering total of $1.39 billion between 1925 and 1928.[2] Chicago's skyline was clear evidence of the city's unprecedented economic prosperity.

Get Your Motors

Running east from Al Capone's Cicero headquarters at the Hawthorne Inn was 22nd Street, a busy thoroughfare that would be renamed Cermak Road following the assassination of Chicago mayor Anton Cermak in 1933. It was there, in an office inside one of the many brownstone buildings lining the street, that W. W. Grainger, Inc., opened its doors in September 1927.

That year, the company distributed three *MotorBook* catalogs. The first of these, No. A, announced a "Bright Future for Dealers Selling Motors" while explaining that "Power companies are rapidly expanding their supply lines in every direction and this means that there are many new prospects for motors."

Factories powered by a single direct current (DC) motor linked to long lines of machines and driven by overhead driveshafts, pulleys, and leather belts started converting to alternating current (AC) motors to drive individual machines, and the whole nation followed suit wherever an electric supply was available. Chief among purchasers were farmers buying motors to operate milking machines, pumps, washing machines, cream separators, feed mills, corn shellers, and wood saws.

"There is nothing seasonable about motor sales," *MotorBook* No. A explained. "It bears a year round, repeat order business." Because they responded to motor shop demand by offering small quantities at low prices, Bill and Margaret Grainger

---

Though motors made up the bulk of Grainger's early sales, the company distributed many other electrical items. The cover of the April/May 1928 *MotorBook* touted fans, calling on buyers to "avoid the summer rush."

had trouble filling all the orders that came pouring in. Joe Svitak and Anne Velkaborsky were hired to help the fledgling company live up to its pledge for quality, low prices, a money-back guarantee, prompt shipment, safe delivery, and service advice. It was just this kind of niche marketing that would attract customers during the Great Depression.

From the outset, Grainger was the exclusive factory sales representative in Chicago for Sunlight, B-Line, and American Motors, with the $\frac{1}{4}$-horsepower Sunlight and $\frac{1}{2}$- and 1-horsepower B-Line motors as its most popular items. These electric motors were strictly factory overruns, but on at least one occasion the company had to buy in bulk from a national retailer in order to satisfy demand. It would take 18 months before manufacturers would accept Grainger orders for bulk production, and so in *MotorBook* No. B, Bill Grainger noted, "Motors are staple articles and we are forced to do a strictly cash business."

Recipe for Success

Increased from 8 to 32 pages, the second catalog offered a noticeably greater selection of items, ranging from drills, lathes, pump jacks, pumps, testing meters, bench saws, grinders, and revolving display tables—even a floor polisher, an air wagon, a ventilator, and the Bee-Vac clothes washer. The front page announced "NEW Motors in Colors" and "This is the day of color and Sunlights are the first to offer you motors in colors."

The much-vaunted colors—red, jade green, and black—would "draw customers to your windows and shop. They will help you sell enough additional through window advertising to pay your rent.... Be the first in your locality to cash in on this distinctively different motor sales stunt at no additional cost whatever to you."

By early 1928, Grainger's catalog listed more than 100 different motors, many already reduced in price due to the company's growing buying power. A $\frac{1}{4}$-horsepower Sunlight motor, for instance, was listed at $8.98, while the $\frac{1}{3}$-horsepower version was $10.35. But because Grainger wasn't yet incorporated, it did not have sufficient credit with manufacturers, a situation that would be remedied just a few months later.

In December 1928, Bill Grainger incorporated the business with a total capital of $6,750, which was provided by his wife, Hally, from her profit-sharing fund at Wagner Electric. That same month his signature could also be seen at the bottom of a statement of intent inside the latest *MotorBook*. Under the heading "Our Platform," the company's founder made a list of promises to customers:

1. *Offer new items first at lowest prices.*
2. *Stock every item so that prompt shipment may be made.*
3. *Ship 95% of orders same day as received.*
4. *Never, knowingly, sell the wrong item for the dealer's purpose, where the application is known.*
5. *Give the benefit of latest prices, if lower, and return balance.*
6. *Refund overpayments promptly.*
7. *Refund promptly when merchandise is returned.*
8. *Guarantee personally every item.*
9. *Answer all inquiries promptly.*

There, in his first written pledge, was Bill Grainger's business philosophy. It would prove a surefire recipe for success. Now viewed by suppliers as a viable enterprise, W. W. Grainger, Inc., experienced continued sales growth along with an ever widening reputation for fast delivery and honest service.

In its early years, the business promoted motors more emphatically than it pushed fans, even though the summer months represented its slowest period. Air conditioning was still a luxury for the very rich, and fans—other than oscillator types measuring up to 16 inches—were not in general use. Thus, no new editions of *MotorBook*s were even published between June and August of 1929.

Still, the April/May edition had expanded to 64 pages and featured catalog numbers for the first time. Previously, ordering had been done by description. Now customers could easily reference products like the Robur Health Massage Machine and its various models, designed to make "exercising pleasant and attractive as well as beneficial."

The Crash

On October 29, 1929, a wave of panic rolled over a sea of dazed brokers, investors, and bankers on Wall Street as the stock market plummeted to an unprecedented and spectacular loss. Within an hour during the frantic day, blue-chip certificates of companies like General Electric, Johns-Manville, and Montgomery Ward tumbled, in some cases losing 25 percent of their value.[3]

For months Chicago was in a state of shock. Many Chicago neighborhoods fell into poverty and suffering. The Great Depression was under way. Yet through it all, the company would continue slow but steady growth. Expansion and diversification would see the company through the worst of times and also create the foundation for Grainger's continuing pattern of success.

The early *MotorBook*s' most popular items included the 1-horsepower B-Line motor (below) and various Sunlight motors, though Grainger offered a growing selection of other products, ranging from sump pumps and drills to the Bee-Vac clothes washer (opposite).

I believe —

The independent merchant is the real backbone of the country.

Factories and consumers should buy motors from the dealer.

The dealer can advise and service them — not just sell them.

The dealer's profit is well earned.

The dealer should be able to meet ruthless direct-to-the-consumer competition.

He should meet it with standard merchandise — not made-to-price merchandise.

The dealer can get more for our merchandise because it is standard.

He can get more because our motors weigh more, will start bigger loads and were built to pull heavy machines — not just for intermittent home use.

If our selling price is regulated, so should the selling price of the dealer's ruthless competition.

We believe that the manufacturer's 17% discount is entirely inadequate for a dealer to properly sell and service motors.

The dealer should buy from us because we are the ones who made it possible for him to make a real profit on motors.

When a dealer buys from the manufacturer instead of us, he hurts himself, because only by bulking his orders with other dealers, through us, can he continue to get the right prices.

The only representative with the manufacturers you have, is Grainger. We fight your battles and stand up for your rights.

We should stick together or our business existence will be at stake.

W. W. Grainger

Even during the Great Depression, Bill Grainger instilled an uncommon bond of trust in his customers with such adages as these, which he published in a 1935 edition of *MotorBook*.

CHAPTER THREE

EXPANSION DURING DEPRESSION

1930–1939

AS THE DECADE OF THE 1930S began, the city of Chicago was broke. The banks refused to extend it credit, and there was no money to pay the city's policemen, firemen, schoolteachers, janitors, or clerks. Unable to pay rent and mortgages, thousands of Chicagoans lost their homes and took to the streets, erecting cardboard shelters and using newspapers and rags for makeshift blankets.[1]

The jobless slept under double-decked Wacker Drive and Michigan Avenue—a spot cynically dubbed the "Hoover Hotel." The Chicago Urban League noted that "every available dry spot and every bench on the west side of Washington Park [was] covered by sleepers."[2] On the South Side, riots broke out as evictions mounted. The number of foreclosures grew from just over 3,000 in 1929 to more than 15,000 in 1933. During the same period, the city's industrial employment was halved, payrolls declined by 75 percent, and those who were fortunate enough to have jobs often received their wages in the form of homemade money called scrip.[3]

Times were tough everywhere, yet Bill Grainger managed to help those around him. David Grainger remembered his father's generosity during one of the most desperate periods of the twentieth century: "I was only six, but I can remember a next-door neighbor coming to my dad in 1933 and saying, 'Bill, if I don't get a job someplace, my children won't eat. Will you give me a job?' And my dad gave him a job. This man worked hard for the company until he retired at age 65."[4]

While many of Chicago's wealthiest businessmen saw their fortunes wiped out during the early years of the Great Depression, W. W. Grainger, Inc., continued to flourish.

In September 1930, the first Grainger-branded motor, a ½-horsepower repulsion-induction model manufactured by Baldor Electric, appeared on the cover of *MotorBook* No. 13. While this particular line would be short lived, it would pave the way for the company's own Dayton brand in years to come.

During the early 1930s, there were no social security taxes, withholding taxes, or minimum wage schedules, and so there was no need to maintain hourly records. However, it is probable that, by 1932, Grainger still had only about 10 employees

---

Bill Grainger recognized from the start of his business that the personal touch goes a long way toward making sales. In each issue of *MotorBook* he included a short, helpful message to the dealer, accompanied by his photo.

on its payroll, including a very young part-time helper who was taking his first baby steps in the family business.

Bank Closings

David Grainger was eight years old when his father started requesting his services. It didn't take young David long to realize how troubled the country's economy was. "When I was in second grade, I worked on Saturday mornings for four hours and got paid 20 cents an hour folding boxes," he remembered. "I put the money in the bank, but then that bank failed, and I couldn't get my money out. The bank paid most of it back, but it took 20 years."[5]

That bank was, in fact, one of at least 163 Chicago banks that collapsed during the first four years of the Depression. Following President Franklin Delano Roosevelt's inauguration on March 4, 1933, the federal government issued the Emergency Banking Relief Act to help restore faith in the banking system. All banks were ordered to close on March 5, 1933, and the stronger ones were allowed to reopen on March 9. In his first radio "fireside chat," on March 12, Roosevelt told Americans that it was safer to "keep your money in a reopened bank than under the mattress." The next day, banks had more deposits than withdrawals.[6] During the four-day "bank holiday," post offices were still operational, and as Grainger had been accumulating postal money orders received in payment of goods, the company was one of very few businesses to have an actual cash flow.

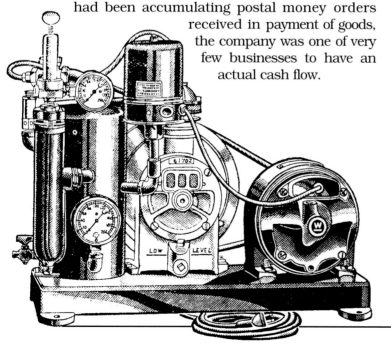

The First Branch

More notable from the company's perspective was another development that took place in November 1933. Knowing that his business needed to get closer to the customer, Bill Grainger initiated a method of expansion that would provide the basic foundation for Grainger's future growth—the local branch office. He chose to open his first branch in Philadelphia, selecting the City of Brotherly Love over the frantically paced New York so that he could test the waters in a less competitive market. Not everyone shared his vision, however.

"My father had a friend, Harry, who was in the refrigeration supply business. Harry and my father were also competitors," David Grainger explained. "My dad said, 'We're going to open a branch in Philadelphia,' and Harry said, 'But [Chicago] is where you live. This is where you know.' My dad went to Philadelphia and Dallas and Atlanta and San Francisco and eventually some 300 other places while Harry didn't open branches, and his business is now defunct."[7]

Bill's sister Margaret was dispatched to Philadelphia as Grainger's very first branch manager.[8] At that time in Grainger's history, the branch manager was everything from manager to phone receptionist to shipping clerk, and Margaret would remain there for more than a year, helping formalize the routines for running a branch. She also helped establish the New York branch before returning to Chicago and assuming the positions of credit manager and company secretary until her retirement in 1953.

New Headquarters

In 1933, Grainger's business volume approached the $250,000 mark. To keep pace with this fast growth, the company required better facilities than those offered by the old loft building

---

A 1933 *MotorBook* featured this "new improved beer pump" just five months before the end of Prohibition. Offering "the best and most economical way of dispensing beer," the pump sold for $42.50.

CHAPTER THREE: EXPANSION DURING DEPRESSION

Above: Unlike many other businesses during the Depression, Grainger was able to expand its operations, opening its first branch in a small corner of this building in Philadelphia in 1933.

Below right: Ed Schmidt, shown here in the 1960s, joined Grainger in 1934 to set up a corporate accounting system. He became president in 1968.

at 700 22nd Street and moved in January 1934 to the third floor of a large public warehouse at 1500 South Western Avenue.

Luxurious compared to the 22nd Street building, the new premises consisted of four standard-sized offices to accommodate the company's 12 employees and a nearby warehouse space of approximately 5,000 square feet. Customers would park outside the main building and take an elevator to the third floor in order to reach the city sales counter, while a freight elevator in another part of the building could be used to transport larger items down to the ground level.

New Talent

Meanwhile, as the tax deadline loomed for 1934 and with his company's records lagging behind the ever expanding business, Bill Grainger sought outside help. A friend put him in contact with the CPA firm of Alexander Grant & Company. This was the beginning of a long and fruitful business relationship with the accounting company as well as one of its partners, Maurice H. Stans. Stans would serve as a close financial advisor to the company over the next four decades and also become a member of the board of directors.

With Maury Stans's assistance, Grainger's 1934 tax returns were filed on time. However, Bill still felt that his bookkeeping department—such as it was—required modernizing, and so Stans suggested that the company recruit the services of one Edward F. Schmidt, who was then a junior accountant at Alexander Grant & Company. Once again the accounting firm's advice led to a long relationship for all concerned. While Bill Grainger always would be the idea man, the one who sometimes moved on a whim, Ed Schmidt, often seen chewing a favorite cigar, would be the conservative check that balanced the company's leadership.

A graduate of the University of Illinois in 1931, Ed Schmidt officially joined Grainger on March 15, 1934, and was immediately charged with setting up a corporate accounting system. As David Grainger told the story, Schmidt went to Bill Grainger after he had finished and said, "Bill, I'm all done." The elder Grainger then said, "Ed, I think maybe you ought to stay."[9]

And stay he did—for almost 40 years. Accounting and finance were undoubtedly Schmidt's primary expertise, but as the company grew he would become involved in all areas of the business, including personnel and legal matters, branch openings, and supplier contracts. Before full-time territory salesmen were employed in 1939, he even called on customers.

Ed Schmidt would become vice president in 1945, executive vice president in 1953, and president upon Bill Grainger's retirement in 1968. Schmidt retired at the end of 1973 and was named vice chairman of the board of directors, retaining this position until his death at age 67 on March 30, 1976.

David Grainger described Schmidt as "a good, savvy kind of person ... not as entrepreneurial as [Bill] but very smart with good people skills, good on protocol. My dad, who never gave speeches, depended on that a lot. Schmidt or someone else always ran the annual meeting, and he was a good sounding board for my father's ideas."[10]

"Ed Schmidt was very conservative in his approach and a very interesting gentleman," noted Lee Flory, who would become a Grainger vice president and secretary. "One of his tenets was 'We're not going to build big branches because the only thing that big branches develop is lots of inventory.'"[11]

Elmer Otto (E. O.) Slavik was another key figure to join the Grainger fold during the pivotal era of the mid-1930s. His family owned Slavik Printing, and at a time when the cost of producing and distributing the *MotorBook* catalog consumed a disproportionately large slice of Grainger's total expenses, Slavik agreed to print it in return for Grainger company stock. With a background in marketing, merchandising, and sales, "He knew how to get catalogs out," said David Grainger.[12]

"It was a mutually beneficial relationship," said James Slavik, who, as a Grainger board member, represents the third generation of Slaviks to be active in the company. "My grandfather left the printing company and became more involved with Grainger."[13]

Known as an outgoing, adventurous man, E. O. Slavik had a strong will and the ability to make things happen. Grandson Jim Slavik once received a letter from a gentleman who remembered E. O., who was then general manager at Grainger, coming into town "with a pocketful of cash" to help the company solve the problems it was having trying to open a branch in the South.[14]

Sharpening the Business

Lessons learned from the early Grainger branches helped the company tailor future branches to maximize their effectiveness. Three new Grainger branches opened in 1934. In May, the Dallas branch opened, serving all of Texas, New Mexico, Arkansas, and Louisiana; the Atlanta branch opened in July; and the San Francisco branch opened in October. The San Francisco branch was the first branch on the West Coast, and for nearly two years this facility would serve the entire region west of the Rockies.

Above: Elmer O. Slavik, whose printing company had produced the *MotorBook* until the 1940s, passed away in 1973. His son, Elmer R. Slavik, and grandson Jim Slavik became members of Grainger's board.

Left: Hard-working, paternalistic, and seemingly always on the go, Bill Grainger is shown here outside the Congress Street building with his prized Lincoln Zephyr automobile. *(Photo circa late 1930s)*

CHAPTER THREE: EXPANSION DURING DEPRESSION

It wasn't long before Grainger became disgruntled with its Atlanta premises. "We have not been completely satisfied with this space since the day the original lease was made," Ed Schmidt complained in a letter to the Atlanta branch manager on December 28, 1934. "The space is not up to par with our other offices, and we are seriously contemplating making a change within the very near future.

"The principle objection we have to our present space is that it is not conducive to city sales," the letter continued. "The customers do not particularly care to enter an old dirty building. There should be a nice amount of business developed right from electrical dealers within 50 miles of Atlanta who will come in person to pick out what they want."[15]

In the same letter, Schmidt also identified the four main factors that had to be considered when selecting a suitable site. These were location, shipping and receiving facilities, safety, and other general features. He went on to explain how the office should be situated as conveniently as possible for dealers, preferably in the wholesale district close to other electrical wholesalers. Proximity to a freight depot and post office was also desirable, while in terms of security, "There should be as few entrances to the space as possible and good facilities for securely locking these entrances."[16]

That same year, Grainger started to offer open account terms as opposed to either cash or C.O.D. Also around this time, the company compiled a new customer mailing list based on certain qualifications. In short, a customer was defined as someone who made total purchases of $5 or more and who had made a purchase within the past year. Furthermore, customers were drawn from several fields: electrical contractors and contractor-dealers; light and power companies; steam and electric railways; federal, state, and municipal governments; large industrial outfits that maintained a permanent maintenance department; manufacturers that purchased electrical materials for incorporation in their products; and established dealers that purchased for resale in the fields of electrical, plumbing and heating, hardware, machinery and engineering, pumps, garages in towns of 2,500 or less, and telephone companies.[17]

Beginning in 1934, Grainger offered open accounts to dealers, demonstrating an uncommon level of trust considering the state of the country's economy.

Others who would be eligible as customers included foundries, farmers' cooperatives, milling companies, theater suppliers, lumber companies, and garages and car dealers in towns with a population over 2,500. However, they were eligible only if, "upon investigation," it could be "proved they buy for resale such equipment as we supply."[18]

The summer months had traditionally been slow for Grainger, but when the company first offered fan parts to customers in 1935, overall sales actually rose during summer. It didn't hurt that the company opened four new branches that year, in Cleveland, New York, Kansas City, and Minneapolis.

As the office in New York was being prepared for its big launch, Ed Schmidt sent a letter to Margaret Grainger advising her how to supervise it from Philadelphia. "We realize it will take a little time to work out definite lines of responsibility between Chicago and New York and Philadelphia and New York inasmuch as this is something of a new angle to our business," he wrote. He advised her to "make free use of the long distance telephone.... We would not be averse to having a call a day."[19]

The Kansas City branch opened in September 1935 and represented an important step in the company's progress because it began the idea of fast counter service and convenient parking. The branch in Kansas City was the first to be located at street level. Direct street access was normally not considered safe, but this branch was within an office building with an entrance accessed from the main lobby. In the future, all new branches would be at street level, and older ones would relocate as soon as possible.

Unlikely Success

During this time, the United States was enduring some of the most difficult years in its history. Yet Grainger continued to expand at a rapid rate, opening six new branches in 1936 alone, more than in any year until 1971. Grainger's office and

As Grainger opened more and more branches around the country, customers were able to enjoy the convenience of quicker availability.

CHAPTER THREE: EXPANSION DURING DEPRESSION

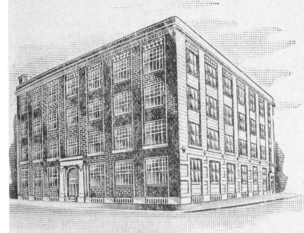

In 1936, Grainger moved its headquarters and main warehouse to a facility on West Congress Street, where it enjoyed four times the space of its previous location.

warehouse space were again being outstripped by product demand and the growing number of customers, and in June 1936 the company's Chicago headquarters and central distribution center relocated once again. This time the destination was a four-story-and-basement building at 819 West Congress Street, where Grainger initially occupied the basement and first two floors, totaling nearly 60,000 square feet. The first fans to bear the Dayton trademark would be manufactured here. Clearly the move had been made with an eye toward the future—Grainger's headquarters wouldn't need to relocate for another eight years.

Optimism ran high at Grainger, and along with opening branches the company continued to diversify its products. July 1936 marked Grainger's entry into the wiring device and refrigeration supply fields, and that same year the *MotorBook* catalog added the General Electric line of motors to its burgeoning offerings.

Grainger's affiliation with General Electric would be a long and profitable one for both companies, but David Grainger recalled that getting the relationship started was problematic. "My father couldn't get General Electric to sell to him originally as a distributor, so he got a friend of his to order a whole bunch of motors," David explained. "The truckload came and my dad bought them from [his friend], and, because it was a whole truckload, there were no labels on them. So all of a sudden my dad is selling General

Electric motors, and finally GE came and said, 'We can only guess where you got them, so I guess we better sell to you.'"[20]

Those who worked with Bill Grainger during the 1930s remembered that he was a master at making deals with suppliers. For example, he struck a deal with GE that allowed Grainger to buy GE motors at a distributor's discount in return for handling GE's smaller manufacturing and dealer purchases.[21]

By 1937 Americans had high expectations and renewed hope for the economy, although much of this optimism would prove unwarranted. There was also hope that America would stay out of the war that was looming ominously in Europe.

With 15 branches, an expanded catalog, and even heavier emphasis placed on the sale of fans, Grainger's overall sales volume was five times that of 1933. *MotorBook* No. 83, issued that February, featured an editorial by Bill Grainger titled "Get Yourself an Umbrella," in which he explained how the manufacturers of more than 1,200 items had initiated price increases of up to 100 percent over 1936 levels. Grainger, on the other hand, had obtained an "umbrella" in the form of a six-month supply of merchandise, thus enabling it to guarantee current prices until stocks were exhausted.

"That is why we are warning you to get yourselves an umbrella in the form of ample stocks," Grainger explained, "so when higher prices rain down, you won't get wet."

That month, Grainger added more than 1,400 new dealer accounts. The year 1937, in fact, still holds the record for the greatest number of *MotorBook* issues published—18 in all. The increased catalog production was a result of Grainger's breaking down its mailing list according to three markets: electrical; refrigeration and heating; and ventilating and plumbing. Individual catalogs were produced for each of these categories. This shift followed a late-1934 attempt to personalize the company by publishing separate *MotorBooks* for all the branches, each with a photo of the branch manager on the cover. In both cases the customized catalogs amounted to a nice idea that didn't work. They were a logistical nightmare, and by the start of the next decade they were abandoned in favor of one standard *MotorBook* for all.

Wade Cautiously

In the meantime, Grainger's leadership was concerned with another burning issue—how to contend with growing competition to Grainger's core business. In a note sent to E. O. Slavik in the fall of 1937, Ed Schmidt identified motors as the company's most popular product and pointed out that competitors in this field had several advantages over Grainger: the repair of old motors, the sale and trade of new ones, and large stocks of used motors. Schmidt recommended that Grainger enter into these fields.[22]

Slavik responded by noting that retailers often advised customers to buy good used motors instead of inexpensive new ones, perhaps hinting that the latter were those stocked by Grainger. However, it would be difficult for the company to change this perception, even by offering a 10 percent discount, as the retailer made a larger profit on used motors.

Slavik thought that Grainger should have "rock bottom prices on the few best-selling motors" and that if the catalog was too expensive to produce, the company should put out flyers that showed its best-sellers several times a year. Slavik also pointed out that, pricewise, some old motors wouldn't be worth repairing. The sale of such items would ruin the company's reputation for exclusively new stock, and besides, it would also necessitate a specialized repair service. On the other hand, mail order sales should help to keep Grainger's prices lower than prices of competitors that used outside sales forces.[23]

"[We] circularize so far and have such sensational things to sell, people will come to us," Slavik asserted. "[We should] limit catalog space to items that pay for themselves, with a certain amount of experimental items.... If [the] catalog is too expensive to be circulated widely, [we should] print an abbreviated catalog or circular just with red hot items."[24]

But reading this, Bill Grainger wasn't impressed with the idea of distributing a circular. He circled the word and wrote a concise and straightforward "no" in the margin.

"Putting the name of a nonexistent company or a skeleton company on the nameplate fosters the idea of lack of dependability," E. O. Slavik continued in his handwritten letter. "Putting the words 'manufactured for W. W. Grainger, Inc., Chicago, offices in principal cities' or something similar, gives stability to the motor.... If we have the service, price, and good product, resale manufacturers could be sold on the fact that we do earn our commission.... [The] most profitable business is that which comes in from small towns, where dealers are not accustomed to having special prices offered to them. They don't ask special favors. They write in their troubles and inquiries, and routine handling can be set up. [I] believe [the] most profitable customer is one who buys about $100 a year. Larger than this, the account tends to become confused and requires expert handling."

Bill agreed with some of Slavik's suggestions but was dubious of others. "I would like to be considerably more confused than we are with $100 annual customers," he wrote. "If we could average $120 per year per customer, our troubles would be over because we could live off our customers. I think you will find as many or more troubles in smaller accounts [as] in larger ones."[25]

Bill Grainger's slight skepticism would prove well founded. Although company sales for the first nine months of 1937 were up 50 percent, the Depression would resume in 1938, and overextended firms would pay the price.

Unfortunately, the company had been misled by its initial success in the fields of wiring material and refrigeration service supplies. Much-needed capital was invested in new stock, and when the bottom quickly fell out of these markets, the inventories had to be dumped at a loss, something which would take the remainder of the year and much of the next to accomplish. The company learned a valuable lesson from the experience: wade cautiously before plunging in too deep.

Belt Tightening

Still, the news wasn't all bad. In January 1938, the Dayton brand name appeared for the first time in *MotorBook* No. 96, gracing a line of three-phase motors. The Dayton Electric Manufacturing Company—Grainger's first foray into manufacturing—officially commenced operations on April 13, and its products would at one point represent the largest percentage of Grainger's total sales volume.

As the decade came to a close, the company was priming itself for an unprecedented program of expansion. In 1939, Grainger began employing outside-territory sales agents for every branch, and shortly this innovation would prove as successful as the branch program itself.

Although no new branches opened between January 1937 and November 1939, Grainger had managed to open 15 during its first 12 years in business. During the next 12 years, Grainger would open another 28 branches around the country.

"When we started out, Branch A was Chicago, B was Philadelphia, and so on," explained David Grainger. "Well, we soon ran out of single letters, so we went to AA, etc., but punch card equipment at that time did not handle letters well, so we switched to numbers. Fortunately we chose three digits instead of two—i.e., we could have 999 branch numbers instead of 99."[26]

The Denver branch, which opened in November 1939, was housed within a tiny, 1,000-square-foot space inside the Sugar Building. It was so crowded that when it was opened up each morning, some goods had to be moved onto the sidewalk before the staff could enter. But the problems in Denver were minor compared to those experienced by employees at the New Orleans branch, which opened that same month. That structure earned the distinction of being the most decrepit branch building. But having to tighten its belt, Grainger was unable to afford better quarters.

Due to losses, including the dumping of its wiring and refrigeration supply lines, Grainger was forced to implement stringent measures. "Because of business conditions, we are sorry to find it necessary to adjust our work-week," announced a March 14, 1939, notice to all employees. "This action has been put off for weeks in the hope that there would be some improvement, but there is now no alternative. Effective as of March 16, the office, warehouse, and fan department will be closed on Saturdays, and there will be a flat 10 percent reduction in salaries all the way through the organization."[27]

A further notice on March 27 helped to clarify Grainger's position. "Losses of over $13,000 in

January and February make it necessary to go to a similar working schedule to that employed last spring or discharge nine employees. Your wish last year was to spread the employment, and we are following the same plan this year. The hour rate will be maintained but hours will be reduced."[28]

As of April 1, 1939, the weekday schedules for office staff would be reduced from 42 hours to 32.5 hours, while those for shipping room staff would be cut from 43.25 hours to 33.75 hours. Saturday work schedules remained the same for all staff.

"We hope conditions will change so you can all be with us, full time, very soon," the memo continued, before adding ominously, "We would appreciate it if those of you who have other means of support, or who can find other employment, would do so. Please cooperate in as good spirit as you showed last year. It is as hard for us to do this, knowing and liking you all as well as we do, as it is for you to accept it. If conditions warrant it, every penny of back pay will be paid you as a bonus later in the year."[29]

On June 1, 1939, full-time hours were restored: 8:30 to 12:00 and 1:00 to 5:00 on weekdays, and 8:30 to 1:00 on Saturdays. However, as of November 17, morning rest breaks were discontinued, while the 10-minute afternoon break was to be temporarily retained.

"There has been an epidemic of absences which has made our record unfavorable with that of other companies," yet another employee memo warned. "A record is being kept of all absences for whatever cause, and deductions will be made from possible bonuses which it may be possible to pay after the close of the year."

Other difficulties plagued Grainger's efforts as well. In October, heating items appeared on the front and back covers of *MotorBook* No. 116. However, as the smallest of these weighed 475 pounds and practically no branch back then had a loading dock, it was virtually impossible to get a stoker off the truck, let alone through the branch doors.

Those catering to the summer heat rather than the winter cold were experiencing no such problems. The late 1930s were big years for fan promoters, who were making good money in the days before it was common for factories, stores, and homes to have in-house air conditioning. Fans were in ever increasing demand, and as high temperatures fueled the rush to buy, the most common method employed by salesmen was

Dayton brand-name motors such as these appeared for the first time in *MotorBook* No. 96 in 1938. Dayton would become among Grainger's most important brands.

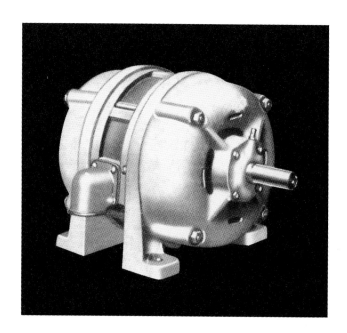
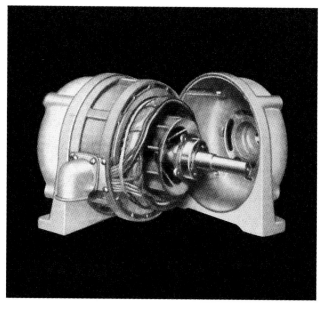

to buy stocks of fans and fan parts from Grainger, assemble them, and then peddle the goods directly from the backs of their trucks. Business in air circulator fans would continue to be brisk for many years.

In September 1939, Britain declared war on Germany, Europe plunged toward the abyss, and, following America's entry into the war a couple of years later, on December 7, 1941, the manufacturing of fans—and countless other products—would be curtailed in compliance with newly implemented war conservation measures.

The challenging decade was giving way to a different, even darker era, yet once again W. W. Grainger, Inc., would turn adversity to its advantage.

Bill Grainger did his part for the war effort by serving as a dollar-a-year man for the War Production Board.

CHAPTER FOUR

THE GRAINGER HOMEFRONT

1940–1945

BY 1939, A TERRIBLE DRAMA OF impending world war was unfolding throughout Europe. The rising menace of conflict dominated newspaper headlines throughout the year: "Czechs Collapse," "Italy and Germany Execute 'Pact of Steel,'" "Nazis Invade Poland," and by the end of September, "Britain and France Declare War on Germany." President Franklin D. Roosevelt had proclaimed America's neutrality in the growing conflict and forbidden the shipment of arms, munitions, and aircraft to any of the countries for which "a state of war unhappily exists."[1] But on November 4, 1939, the embargo was lifted, and the United States began a massive buildup to supply the beleaguered nations of Great Britain and France with armaments and the many goods of war.

In 1940, a *Chicago Tribune* statewide poll revealed that 74 percent of those questioned preferred isolationism to entering the growing conflict in Europe.[2] A steady stream of antiwar *Tribune* editorials underscored the view that the nation was both ill equipped and ill advised to partake in a conflict that was very much "over there," and the Chicago committee to keep America out of the fighting contrived to have aviation hero Charles Lindbergh reiterate these points to a crowd of 40,000 at Soldier Field on August 4, 1940. While *Chicago Tribune* owner Colonel Robert McCormick spoke out bitterly against the draft law that was passed by the U.S. government in September 1940, he was strongly opposed by supporters of President Roosevelt, and on December 4, 1941, Chicago interventionists at last found a voice in the city's new prowar daily newspaper, Marshall Field III's *Sun*.[3]

Then on December 7, 1941, over 360 Japanese warplanes participated in a disastrous attack on Pearl Harbor, sinking or very seriously damaging five U.S. battleships and 14 smaller ships. Over 2,000 Navy personnel perished, along with 400 civilians. President Roosevelt called the attack "a brilliant feat of deception, perfectly timed and executed with great skill." Within four days of the attack, the United States would declare war on Japan and its Axis partners, Italy and Germany.

Once America joined the horrible fury of world war, the antiwar committees dissolved, the *Tribune* did an about-face, many isolationists went into uniform, and pictures of Adolf Hitler disappeared from the walls of the German bars and restaurants that lined Milwaukee Avenue. Chicago soon led American cities with its $1.3 billion expenditure on war production plants capable of mass-producing airplane engines and other emergency

---

The bolts of lightning on the Grainger logo of the 1940s illustrated the company's reputation for speed.

Ed Schmidt (far right foreground) created many of the Chicago office's policies regarding hours, compensation, accounting, and general operations. *(Photo circa 1946)*

items. Chicago also contributed on an altogether different scale when a group of scientists led by Dr. Enrico Fermi managed to split the atom; just after 3:30 P.M. on December 2, 1942, in a secret laboratory under an old football stadium at the University of Chicago, the nuclear age was born.[4]

Employee Benefits

The ration-conscious war years would deprive American factories, stores, and citizens of numerous goods and services, yet during the summer months of 1940, few thought that the faraway conflict would actually cause scarcity and increased prices within the United States. Bill Grainger, Ed Schmidt, and E. O. Slavik were among the few.

Even though sales for the first six months of 1940 were up by more than 20 percent over the first half of 1939, management was aware of the need to build up inventories quickly, before government cutbacks were announced.

An October 1940 amendment to the U.S. Wages and Hours Act reduced the maximum 42-hour workweek to 40 hours. The change would have resulted in reduced wages for Grainger employees, but the company made sure that its employees worked just enough overtime (one hour and 20 minutes) to match what they would have made for 42 hours. Grainger's workforce would receive the same weekly wage for less time worked.

"The wages of a person doing desk work should not be judged by hours but by work accomplished," Ed Schmidt suggested in a handwritten note to his colleagues on October 19, 1940. "They do not necessarily go together. A person ... can easily do his work in 15 minutes less per day, or even much less, if in [the] right frame of mind. On [the] other hand, he could take 2 hours longer if not feeling exactly right. The differential between 40 and 42 hours working time will not have a material effect on the amount of work turned out."[5]

There were other ways in which the company looked after its employees. Grainger gave its workers a 10-minute rest period at three o'clock each afternoon "to give all employees a given time during which to visit, relax, etc." There were bonuses paid after the first of each year, the total of which could run as high as 20 percent of Grainger's earnings during the preceding 12-month period. Each employee's share of the total was based on the following criteria (in this specific order): general attitude, efficiency, attendance, and years of service. A typical bonus handed out at the start of 1940 was $10 minus social security tax, and an equivalent amount would be a Christmas bonus at the end of the year.

After six months of service, each employee also received $500 of life insurance with AETNA Life Insurance Company, which would increase by $100 a year up to a maximum of $1,500, with all premiums paid by Grainger. Two weeks' paid vacation was also provided to employees with at least one year of full-time service, while one week was given to those who had worked for six months. Sick leave was paid for a maximum of six weeks, small loans without interest or carrying charges were advanced in the event of emergencies, and in December 1940, the company provided a hospitalization insurance policy for all employees with at least six months of service. The benefit rate was $2 per day while hospitalized, a sufficient amount at the time, plus various amounts for surgical bills, as specified in the policy. Again, Grainger paid the premiums.

Hodgepodge

In the meantime, by March 1941 some effects of the European war were being felt in America, and the nation's business leaders were beginning to plan ahead. Clearly, many items would be in short supply, and Bill Grainger's editorial in *MotorBook* No. 133 warned that "the dealer with stocks of goods will get the lion's share of the business."

Due to its planned stockpiling, Grainger's inventories were sufficient to meet the challenge. During the coming months, its distribution system would be relied upon to deliver strategic material where most needed. However, by April, with supplies already tightening, a special flyleaf attached to the cover of *MotorBook* read, "Prices up as much as 16%. But we will protect you until May 1, 1941."

By May, Grainger was already turning its attention to marketing less-routine items, such as the Master Portable Milker machine in that month's *MotorBook*. Priced to sell at $125, it could milk two cows at once, powered by either an electric motor or a gas engine.

That July, the title of Bill Grainger's *MotorBook* editorial was "Food For Thought." Grainger referred to the tons of surplus goods that customers had offered and which W. W. Grainger, Inc., had bought and resold. "Let me know personally if you make something we can sell in our catalog, and if we can help you dispose of surplus stock," he stated.

It was also around this time that Bill Grainger had plenty of other food for thought—quite literally.

Billie's Chili

For years Bill had enjoyed Royal Chili—so named because it was created by Royal Ritchey,

During the lean World War II years, Grainger offered an eclectic mix of nonpriority items such as the Master Portable Milker, which sold for $125 and could handle two cows per milking cycle.

Hally's cousin, who sold it in a restaurant located next to Harry Truman's law office in Independence, Missouri. (Truman also enjoyed Royal Chili.) Bill finally succumbed to temptation, shelled out a whopping $5,000 for the recipe, and set about launching the American Chili Company.

"After my father died, I went through his things and found the recipe in every drawer," said David Grainger. "There must have been 15 or 20 copies, just to make sure it didn't get lost. I've had it made since and it's really good, but it's very expensive. There's no filler, it's all meat, and you have to make at least 50 pounds to get the flavor right. Whenever I've had some chefs make it, my wife has said, 'Oh, honey, another 50 pounds of chili?!' But it's only lasted a week because everybody wanted some."[6]

Allocating himself 751 shares, Bill enlisted partners in the American Chili Company: his wife, Hally, who had 249 shares; E. O. Slavik with 200 shares plus another 180 for serving as trustee; Ed Schmidt with 60 shares; Antoinette Slavik with 50 shares; and Maury Stans with 10 shares. In November 1940, they secured space at 13 South Crawford Avenue in Chicago and, trading as Billie's Chili, tried to hook the great Chicago public on the much-vaunted recipe. Chicagoans are noted for their hearty appetites, but in this case they evidently didn't bite.

Cush Bissell (founding partner of Lord, Bissell, and Brook), Ed Schmidt, Bill Grainger, and Maury Stans (of Alexander Grant & Company) at the Edgewater Hi-Jinx golf tournament in 1944. According to son David, Bill would "come home from work, roll up his sleeves, and play golf in his white shirt and suit pants."

"I was in eighth grade when Billie's Chili was operating," David Grainger recalled. "A friend's older brother worked there on Saturdays, and it was just a golden place to go. However, my mother didn't like it. She told my dad to pay attention to the other business, so it didn't last long. You've got to be a certain kind to make a restaurant go, and this was not going to happen."[7]

Chicagoans were more interested in feasting on other local favorites, such as shrimp dejonghe, chicken vesuvio, the Italian beef sandwich, and within a few years the deep-dish pizza. The American Chili Company went out of business in July 1941, and "Billie" and his partners refocused their attention on their other, far more prosperous enterprise.

The Profit Sharing Plan

That year, Grainger inaugurated a revolutionary incentive plan that to this day best demonstrates the company's concern for the welfare of its workforce—the W. W. Grainger, Inc. Employees Profit Sharing Plan. Grainger would contribute part of its net profits each year into a profit sharing trust (PST) fund set aside for its employees. The contribution was then allocated to eligible employees' accounts based on a combination of their years of service and annual wages. The account could be paid out once an employee became eligible to retire. Most Grainger employees have been able to retire quite comfortably as a result of the plan, and more than a few have retired as millionaires.

Although not entirely unique even in the early 1940s, Grainger's Profit Sharing Plan was certainly the exception rather than the rule in terms of company-provided employee benefits. Fewer than 400 such plans were in existence throughout the United States, and the Sears Plan differed from Grainger's in that it required employees to contribute 5 percent of their salaries for the company to match.

"In line with our policy to build our business toward greater security for all our loyal employees, we wish to announce the creation of a $39,000 profit sharing fund," the company announced. "This costs you nothing, the company makes all contributions. In 1941 your share equals about 18 percent of your total wages, including bonus."

"It's enforced savings that the employees don't have to fund themselves, and it's tax-deferred," explained David Grainger. "So instead of being eroded away by taxes, the full boat gets invested for you. After a year you get a 20 percent share, and then a 40 percent and a 60 percent and an 80 percent and then 100 percent after five years."[8]

As Grainger approached its 75th anniversary in 2002, it was agreed that the Profit Sharing Plan probably did more to evoke employee loyalty than any other single factor, for when employees are actually rewarded for how well the company is doing, they are more likely to devote themselves to their jobs. "I remember working as a sales manager down East for two years back in the late fifties. When I'd explain the scheme to new salesmen, they couldn't believe it," recalled David Grainger. "The Profit Sharing Plan really has been an incentive to stay with the company."[9]

"The Profit Sharing Plan has withstood the test of time," said Nancy Thurber, director of benefits. "It has become such a cultural thing for Grainger. When you alert employees that things change and we need to look at costs, the biggest cry you hear is 'Don't touch the PST!'"[10]

But perhaps Ed Bender, who began working at Grainger in 1959 and retired in 1998 as a vice president, summed up the merits of the plan best when he said, "When you get up to 20 percent of your income put into a trust fund over and above your income paid out to you, that's a hell of a savings plan."[11]

Shifting Priorities

Meanwhile, by September 1941, the nation was truly in a state of emergency. With new products no longer available, *MotorBook* replaced the usual new lines for the fall and winter with policy statements regarding defense orders and pricing. Despite the shortages, Grainger maintained a comprehensive building program, with new branch buildings planned for Cleveland, Miami, and Buffalo while others were under construction in Kansas City, New Orleans, Providence, Cincinnati, and Pittsburgh.

America's entry into World War II that December prompted the nation's industry to convert to all-out production of strategic materials. Electric motors and production-related equipment were vital to the war effort and available only in accordance with the War Production Board's priority system, meaning that they were rated according to their scarcity while businesses were rated according to their products' importance. *MotorBook* No. 145 listed the priority ratings required to buy products from Grainger, ranging from blowers, speed reducers, integral horsepower motors, motor switches, and pumps—which were all very scarce—to the unit heaters that could be sold only with a specific release from the War Production Board.

"We couldn't sell anything without some kind of priority," said David Grainger, "because otherwise we couldn't replace it. But what we had, we could sell, and we sold all kinds of stuff: toys, radios, Swiss watches, even World War I wagon wheels we found in an Army depot and copies of the New Testament with steel covers for soldiers to wear over their hearts to protect them from shrapnel."[12]

For Grainger, which provided an important distribution function in the maintenance-and-repair priority category, the next four years would be unlike all others in the company's history. Material shortages in its lines wouldn't become really acute until late 1942, and in January of that year, it followed through with its plans to open the Miami and Buffalo branches. February's *MotorBook* announced that immediate delivery would be provided for motor orders accompanied by an "A10" priority. Grainger was obtaining motors from appliance manufacturers who were converting to war distribution, and it was then able to distribute the motors to manufacturers who were in need of them. Orders for fans were also on the rise due to many customers' belief that they would soon be in short supply. "If you had a motor on the counter during the war, you could sell it in an instant," said David Grainger. "Any kind of motor would do. You could not put motors where anybody could see them by looking into the warehouse because people would come around the counters and take them."[13]

Grainger was challenged by the war in other ways as well. In January 1942, the company began demanding "preference rating certificates" for all orders relating to national defense items. By March, some of Grainger's employees were being drafted, and paper shortages meant that the demand for *MotorBooks* could no longer be met. "Conserve this catalog" was the message on the cover of No. 146.

But Grainger was never the type of business to allow itself to be overtaken by events. On the contrary, much of its strength was derived from its ability to evolve and adapt to changing circumstances. *MotorBook* No. 148 featured two pages devoted to "blackout items" such as air-raid sirens, blackout lamps, and candle holders, and during the summer of 1942 Bill Grainger adopted a timely and patriotic stance in order to promote the sale of fans through the catalog. After all, Grainger pointed out, not only would the fans aid working conditions, but by improving the health and morale of defense workers, the fans would boost the production levels of companies involved in the war effort.

Generous Appreciation

At Bill Grainger's suggestion, there had been a revival of the regular department-head meetings that had been held several years earlier. "Every alternate Monday we will dine at some suitable restaurant, in a private dining room, and for one hour after lunch, discuss one subject important to the progress of our business," E. O. Slavik informed his colleagues in a memo on March 2. "Everyone

Among the wartime items Grainger sold was the Shields of Faith Bible, which "fits the pocket over the heart to protect our loved ones spiritually and to deflect bullets, shrapnel and bayonets that may endanger loved ones."

will be expected to take part in all discussions as a condition to continued participation. Discussions will be fact finding rather than critical and for the purpose of increased coordination."

The main topic of discussion for the inaugural meeting, held on March 9: "Methods of assuring 1941 volume in 1942." Grainger met this particular goal with room to spare. Yet despite the increased productivity, the federal government's Wages and Hours Act made it impossible for Grainger to pay its workforce larger bonus checks that Christmas than it had distributed the year before unless it received special permission from the Stabilization Board and the Treasury Department.

"We want to pay you a substantial amount more than that, and we have applied for permission to do so," a company memo stated. "If our application is denied, we expect to find other means of legally increasing your compensation, and we hope you trust our sincerity in this matter."

The company's application was approved in January 1943, and supplemental bonuses were paid to all employees shortly thereafter. Faith and effort on both sides had been justly rewarded, and it was in this same spirit of cooperation that, just a short time earlier, the Grainger hierarchy had received a signed thank-you note from the workforce, "imbued with deep appreciation and as a means of expressing our sincere fidelity to our employers for the many generosities bestowed now as in the past."

The note went on in a highly formal tone:

We, the employees of W. W. Grainger, Incorporated, herewith affix our signatures, and in aggregate effort extend enthusiastically our thanks for the companies [sic] latest policy in adoption of the staggered hour working week, thus eliminating Saturdays from the regular schedule as heretofore. We feel that this not only is a patriotic gesture in full cooperation with the Government's transportation conservation program, but also benefits immeasurably each of us as individuals. We are and shall ever be mindful of this in discharging our assigned duties, or in whatever other way possible.

There could be no doubting the employees' sincerity, enthusiasm, and gratitude, but a motley bunch of slackers temporarily spoiled the party.

On February 15, 1943, a memo from E. O. Slavik to all employees effectively warned against taking advantage of Grainger's munificence.

Your company has attempted to exercise as little restraint over you as is possible without harm to the business. In the past month some of our people have been tardy from 3 to 25 times. This is unfair and injurious to the welfare of us all, and makes it necessary to announce that starting Feb. 1st your free Saturday privilege will be taken away to the extent of one Saturday for each tardiness in excess of 2 per month. Such a Saturday will be in addition to any Saturdays spent in keeping our work up to date. If tardiness persists it may be necessary to start the work day 30 minutes later, with the day ending at 5:30 pm. We hope this notice will remind the tardy that they are throwing extra burden on others in their department— a practice that is definitely unfair.[14]

Tailoring Stock

In March 1943, food rationing began. There were restrictions on meat and canned goods, with the government promoting "meatless Tuesdays" as vital to the war effort. But food was far from being the only commodity in short supply. Like everyone else, Grainger was now curtailed by low or no stocks of its regular merchandise, and so it continued to make available to its customers an ever wider array of unconventional items. That May, for instance, the back cover of *MotorBook* No. 160 advertised a coffee grinder and dinnerware, while No. 162 listed chinaware, table lamps, world globes, mirrors, pictures, dresser sets, pen and pencil sets, cookbooks, toys and games, playing cards, encyclopedias, and wooden wagons. In April 1945, *MotorBook* No. 183 even listed an infrared lamp chick brooder, a "no-priority-required" item selling for under $9 and capable of handling up to 300 chicks.

Due to the priority status of practically all metal items, wood was being used for numerous appliances. Grainger offered wooden fan blades, for example, delivered immediately and with no priority requirement. One notable exception to the no-metal-items regulation was the all-metal smoking stand that appeared in *MotorBook* No. 175 at a bargain price of $6.45 for a lot of 10.

Meanwhile, in June 1943, the War Production Board disrupted one of Grainger's key markets when it ordered the company to freeze its supply of fans and air circulators. For the time being, these could be released only upon the receipt of specific instructions from the War Production Board. To purchase them, the user had to fill out application form PD556, explaining how these items were to be used in relation to the war effort. The form was passed to the dealer, then to Grainger, and then on to Washington.

Red tape became the order of the day. Yet while the company had to contend with restrictions and a growing avalanche of paperwork, it continually simplified transactions by ensuring that customers received what they ordered. A 1943 note from Ed Schmidt to E. O. Slavik made the following suggestions:

1. *Eliminate as many substitutions as possible. On non-priority items where the merchandise would probably sell to a customer who specifically ordered same, do not substitute such items on other orders, but cancel.*
2. *Hold on backorder only high priority orders where merchandise is definitely known to be coming in.*
3. *Cancel all non-priority backorders since goods will probably sell anyway.*
4. *Pack merchandise better to eliminate breakage.*
5. *[Use] as realistic advertising copy as possible to be sure dealer really gets what he expected.*
6. *Take merchandise out of catalog as soon as stock is exhausted, or when nearly exhausted if leftovers can be disposed of by other means.*[15]

Company Security

By September 1943, as the tide of the war began to turn in the Allies' favor, forward-thinking businessmen were already sending in orders for "postwar delivery dates." In fact, the company was so inundated with requests that it soon had to set up a special postwar sales department to handle all the orders. As things turned out, this was somewhat premature, but a year later, in his *MotorBook* editorial, Bill Grainger would be urging dealers to order consumer goods for postwar delivery at 1941 prices, and subsequent events would prove this to be sound advice. During the postwar boom, prices would skyrocket as soon as the goods rolled off the production line.

Meanwhile, in January 1944, following an order by the War Manpower Commission to establish a 48-hour workweek, Ed Schmidt sent a memo to

The Infra-Red Lamp Brooder was a "no-priority" item, and *MotorBook* urged dealers to "put this quick-turnover item in your line now."

all employees in which he pointed out that most of Grainger's workers were already working more than 48 hours and that those people who weren't would now be obliged to do so. All work in excess of 40 hours was to be paid for at a rate of time-and-a-half.

That September, in response to the War Production Board's approving up to 5 percent additional salary through payments of employee insurance, Grainger increased (and at least doubled in most cases) its employees' life insurance and hospitalization benefits while adding hospitalization coverage for dependents. In short, employees would benefit from $1,500 of life insurance, hospital room expenses up to $6 per day, surgeon's fees up to $150, operating room and x-ray expenses up to $30, and the hospital room expenses of dependents up to $3 per day.

"We are glad to be able to give you these additional benefits," the notice to employees concluded, "because it further adds to the security it is our ambition that our company provide for us all."

But nothing that the company did could provide the level of security that people began to feel when war in Europe finally ended. Grainger closed its doors in celebration of V-E (Victory in Europe) Day on Tuesday, May 8, 1945, and again in early September

Grainger offered this "new, unused" gas mask and bag only to selected dealers in surplus material.

when Japan's official surrender finally ended World War II. All priorities were abruptly canceled, as were production restrictions on civilian goods. After nearly four years of conflict in foreign lands, America's troops were coming home. It was a time of rejoicing and reacquaintance. But for W. W. Grainger, Inc., the new challenge was to acquire new merchandise.

GRAINGER GRAPEVINE

December, 1948

MERRY CHRISTMAS TO EVERYONE

This is the first issue of "The Grainger Grapevine" -- a paper for and about all the Grainger gang. We hope you will help to run it, with your ideas and suggestions, but most of all, with news about you!

Every publication has an aim or goal. Ours is to promote a unified feeling among all members of the company. We want to tell something about each employee, so that everyone will feel kinship with, find something in common with, someone who was perhaps only a name or face before.

The response to our questionnaire was very gratifying, for it was obvious that those of you who answered had really thought about the questions and about the name of the paper. It was apparent, too, that the idea of having a new house organ met with great approval. We hope we can live up to expectations, and that your enthusiasm will continue.

Your support is essential to the success of this enterprise -- so please, those of you who have not yet returned your questionnaires, take time out and tell us about yourself -- we'd like to know you. And send them back quick!

ORCHIDS TO MRS. COPELAND

Our thanks go to Mrs. Constance Copeland, of Chicago, for the clever name she suggested for this publication. Many good names were submitted, and it was difficult for the judges to narrow down the choice. So thanks, everybody, for your entries, and congratulations, Mrs. Copeland.

MR. SLAVIK RETIRES

Fulfilling a long felt desire, Elmer Slavik has retired from active participation in Company management, and will move with his family to California. Mr. Slavik retains his financial interest in the company and will continue to be available on a consulting basis. In California, he expects to enter some type of business and we all wish him the greatest possible success. Management of our company will continue in the hands of Bill Grainger and Ed Schmidt.

GRAINGERS RETURN FROM EUROPE

Mr. and Mrs. Grainger returned to Chicago on Nov. 4 from a trip through Europe. They sailed on the Nieuw Amsterdam and were met in London by the Frank Vavres (Chgo) who were also vacationing abroad.

In Glasgow, the Graingers visited several of Bill's cousins whom he had never seen, and met Mrs. Susan Stewart, mother of Mrs. Cathy Keenan (Chgo). They also visited John Halliday's (Chgo) family in Dumfries, Scotland. In Oslo, Norway, the Graingers enjoyed meeting the sister of Clara and Jennie Jensen of Chicago, although they had to talk through an interpreter.

They were in Holland in time to see a festival in a small Dutch town, with all the trimmings -- wooden shoes included. Mr. Droeste escorted them through the famous chocolate factory. They ended the trip in Paris, where, of course, Mrs. Grainger bought a very attractive hat. Both agree that it is wonderful to be home in the U.S.

The premiere issue of the *Grainger Grapevine*, in December 1948, served its purpose well: "to promote a unified feeling among all members of the company." The first editor of the *Grapevine* was Bill and Hally Grainger's daughter, Barbara.

CHAPTER FIVE

SCARCITY AND BOOM

1946–1949

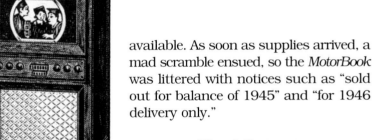

THE RETURN OF VICTORIOUS troops to Chicago after World War II prompted citywide celebrations, a great deal of emotion, and a steady supply of new jobs waiting to be filled. The war had reversed the economic effects of the Great Depression, and the city, like the rest of the nation, was prospering. During the next few years, Chicago's population would rise by more than 230,000, and while the war had eradicated Chicago's unemployment problems, in many areas the streets were still filthy and corruption was widespread, from City Hall officials to police chiefs and officers.[1] The mob's brothels and numbers racket flourished on the South Side, nearly every bar had its own dice game, bookmakers operated out of regular storefronts, and Chicagoans routinely flashed five- or ten-dollar bills along with their drivers' licenses when stopped for traffic violations. There was a reason comedian Mort Sahl referred to the city's Outer Drive as "the last outpost of collective bargaining."[2]

For its part, W. W. Grainger, Inc., experienced rapid growth during the postwar boom years, but while industry was still realigning itself for peacetime, the skyrocketing consumer demand often far outstripped the company's supply of goods. Dealers wanted fans and fan accessories, and they wanted them immediately, but Grainger had to delay filling orders until materials became available. As soon as supplies arrived, a mad scramble ensued, so the *MotorBook* was littered with notices such as "sold out for balance of 1945" and "for 1946 delivery only."

Wage Adjustments

After the war, it made sense for Grainger to make some adjustments in its payroll policy, especially following the abolition of wage controls in September 1945. Not only did the extra pay help employees cope with the rising cost of living, but it also attracted new employees. The minimum salary for sales representatives was increased to $200 per month, and the indoor branch employees' minimum wage was adjusted to 75 cents per hour, while office personnel saw their hourly minimum rise to 60 cents. In the shipping room, the minimum was about 65 cents per hour, and among administrative staff the monthly minimum rose from $200 to $250.[3]

Ed Schmidt proposed performance-based compensation as well. "Award merit increases in hourly rate once or twice each year to deserving

---

During the postwar years, Americans were clamoring for luxurious household items such as this "Trav-ler Television," which Grainger sold in 1949 for $198.98.

Chicago girls and shipping room help, based on ability and seniority," he suggested. "Those who do not get [a] raise will work for one next time."[4]

Furthermore, Schmidt proposed abandoning the prepaid overtime system. "Under our bonus system (in most cases bonuses will still be substantial) prepaid overtime is no longer necessary and is the source of unnecessary ill-feeling," he stated. "Pay employees on holidays or when sick for hours worked. If we reach a bad year when we can't afford this, we can always revert."[5]

The Grainger Policy

In October 1945, E. O. Slavik authored a company strategy entitled "Résumé of Grainger Policy." In short, the document reiterated the need for product diversification as expressed by Bill Grainger in *MotorBook* No. 2 shortly after the company's inception, but it also stressed the need for market diversification through the branches. It also emphasized the need to learn from mistakes, the importance of never gambling on inventory, and the recognition that facts should determine new sale items. Finally, Slavik's policy set 70 branches as the company goal, along with the eventual purchase or start of a motor factory.

Several of the points made by Slavik would guide Grainger for years to follow:

In 1929, [a] major policy manifested itself with the below-market purchase and highly profitable sale of General Electric Repulsion Induction surplus Motors.... Surplus purchases should always be used to swell our margin of profit....

In 1936 ... we made a set-up with the Continental-Illinois Bank for a $50,000 line of credit which today has become a $250,000 line. This has been one of our greatest sources of profit. If you can borrow at 3% and make 5% on your borrowings you gain 2%. We should always borrow as much as we can profitably use.

We should NOT advertise a manufacturer's name where such a name does not command sales— USE OUR OWN NAME.

In judging new items or new actions, we should always use our records rather than our judgement. Dig up old Kardex, or sales, or profit records on the nearest similar deal and be guided by that rather than your own inclinations.

In buying we should always buy at the quantity that affords the best price because, while expenses are important, a 20% extra discount would pay all of our expenses. A 1% saving in prices is 4 times as good as a 1% cut in expenses. A 10% increase in sales is easier to get and produces as much as a 10% cut in expenses.[6]

Many years later, on February 22, 1991, Richard Keyser, who became Grainger's president in 1991, circulated the policy statement to fellow executives Wiley Caldwell, David Grainger, and Jere Fluno, noting, "There is a lot of wisdom in this policy that is useful to bear in mind as we go forward."

Meeting Consumer Demand

Meanwhile, during the last quarter of 1945, while four new Grainger branches, in Oklahoma City, Portland, Memphis, and Toledo, brought the total number of branches to 30, *MotorBook* items were still priced in accordance with government ceilings in order to combat soaring prices.

The following February, the cover of *MotorBook* No. 191 announced in large, bold type that all single-phase motors and all types of electric fans were already sold out for the year. Manufacturers had been making every effort to meet consumer demand, but they just couldn't keep up. "Sold out into 1947" and "indefinite delivery" became commonplace catalog announcements. "There wasn't much in the catalog in 1946," said Bob Wiggins, who became a divisional sales manager that year

---

Fans were in short supply immediately following the end of World War II, and the cover of *MotorBook* No. 191 in 1946 announced that all electric fans were sold out for the entire year (opposite). By the end of the decade, however, Grainger was able to offer a variety of fans, such as the 16-inch adapter fan unit above.

CHAPTER FIVE: SCARCITY AND BOOM

and later became vice president of sales. "We had blowers. We had pumps, and whenever we had fan blades, all we had to do was say we had them and we sold them."[7]

At the same time, factories were selling to Grainger based on price at the time of shipment, forcing Grainger to follow the same practice when shipping its orders to customers. Bill Grainger described this method of doing business as both unnatural and out of keeping with the American way of life. He appropriately titled his May 1946 editorial "Starving amidst Plenty."

That June, production and demand for fans had escalated so high that Grainger's biggest handicap became a lack of cartons to ship them in. As a result, fans and fan parts were offered to resellers who could provide their own means of picking them up. Dealers rented or borrowed cars, trucks, and even moving vans, which arrived in Chicago from as far as New York and Philadelphia. And, as all sales were cash only, it wasn't unusual for dealers to carry thousands of dollars as they drove across the country.

By July 1946, the order file was bulging for small household appliances, and Grainger shipped items on a first-come, first-served basis. In September, the company reported that production and sales of over-the-counter consumer goods had reached record highs during the preceding six months, and by the end of the year, with shortages easing, Grainger began relying on impending price decontrols to stimulate production and deliveries. Still, it would take another four months for the supply of fans and fan parts to meet demand, at which point Grainger had around 50,000 customers.

With the economy flourishing (in direct contradiction of many economists' predictions), the surplus consumer demand for commodities such as cars and refrigerators meant that Grainger's biggest problem was price stability. High retooling costs to convert to consumer production, combined with the public's buying spree, drove prices up. Consequently, Grainger selected its suppliers carefully, buying mainly from those who agreed to protect prices. But for all of Grainger's efforts, some price increases still had to be passed on to customers, though the company made it a policy to pass along price reductions from manufacturers as well.

Around this time, Grainger made its first foray into international distribution with branches in Havana, Cuba; Mexico City, Mexico; and Rio de Janeiro, Brazil. Unfortunately, the branches in Latin America proved unprofitable, and in 1947 David Grainger worked with his father to close them. Years later David noted that Grainger's early international efforts didn't work out because it was "politically impossible to be profitable" in those locations at that time. Not until the 1990s would the company again open branches in Latin America, in Puerto Rico and Mexico.[8]

The Grainger Gang

It took a few years following the war for production to catch up to consumer demand, but the year 1948 heralded the start of a golden era for W. W. Grainger, Inc. By that time, the profit

> # EMPLOYEES TOSSED TOGETHER
>
> IN JULY 1946, MANAGEMENT HESITATED before formulating plans for the Chicago office's annual summer picnic. A special four-person committee was appointed to decide whether to take a summer outing at all and, if so, what form it should take. The indecision resulted from a novel idea the previous year that had turned into a near disaster. Known in company folklore as "The Famous Chicago Boat Trip of 1945," it evolved from a well-intended decision to take a pleasure ride across Lake Michigan to Benton Harbor.
>
> Saturday, July 14, 1945, was gray, cool, and windy, and as soon as the ship *City of Grand Rapids* departed from the Michigan Avenue bridge at 9:45 in the morning, it began pitching and tossing. A few staterooms were hastily requisitioned, and people started piling in, two and three to a bunk. When lunch was served, only about half of the 75 passengers had the stomach to eat anything, so it wasn't surprising that a dozen of them remained skeptical regarding the captain's announcement just prior to arriving at Benton Harbor that calm weather had been forecast for the return trip. These wise souls opted to take the train home instead. As events would soon prove, their instincts were sound.
>
> Everyone who boarded the ship for the return voyage that evening quickly discovered that the sea was rougher than ever. After plates and food began flying around the kitchen, the dining room crew gave up on serving dinner, not that many passengers were disappointed, and the cashier even deserted the register.
>
> The tradition of annual summer picnics would continue, but the Chicago office never launched another boat trip. And as for the good ship *City of Grand Rapids*, it went out of service about a year later, along with company excursions to Benton Harbor.

sharing trust totaled nearly $447,000, and sales volume had increased 25 percent over the preceding 12 months to a total of around $8 million.

In December 1948, Grainger produced the first issue of its company newsletter, the *Grainger Grapevine*. Consisting of three mimeographed pages of pink paper stapled together, the newsletter described itself as "a paper for and about all the Grainger gang." Its self-proclaimed aim was "to promote a unified feeling among all members of the company." Although the *Grainger Grapevine*'s format and appearance would continually evolve, the name of the newsletter and its goals would remain the same for the next 34 years, until it underwent a complete redesign and was retitled *Network* in January 1983.

As had been Grainger's tradition, the company sought input from its employees about what sort of content the company newsletter should have and what it should be named. The first issue, edited by Bill and Hally Grainger's daughter, Barbara, included a report about her parents' European vacation, a profile of Los Angeles branch manager Jack Streeter, and various items of "chatter."

That same issue also announced the retirement of Elmer Otto (E. O.) Slavik, who had been with the company since the mid-1930s, when he helped to produce and distribute the first of many *MotorBooks*. Slavik had been an integral part of the company during the Great Depression and the tumultuous years of World War II. "He was a very flamboyant, outgoing type of guy," David Grainger recalled. "E. O. had been good for the company, but at times he was too adventuresome for my parents' way of thinking. Often my father and E. O. saw things differently. Eventually it didn't work out, and in 1948 the Slavik family left for California."[9]

E. O. "loved to experience a lot of different things and didn't want to leave any experiences undone in his life," added James Slavik, grandson of E. O. Slavik. "He was a fun-loving guy, an interesting guy, but he was also very strong willed."[10]

CHAPTER FIVE: SCARCITY AND BOOM

Eventually one of E. O.'s two sons, Elmer Richard (E. R.), would be elected to the Grainger board, and the elder Slavik would even return to the board for a short time at the start of the 1970s, when E. R. was experiencing health problems. James Slavik, E. R.'s son, would become a Grainger board member in 1987.

Improving Efficiency

By 1949, Grainger had added the Westinghouse line to its prestigious assortment of brands, and the company consistently set monthly sales records. Still, net profits were down compared to the previous two years, despite the guidelines set down in its Guide to Purchasing dated January 18, 1949:

We are interested in selling anything we can just so long as there is a net profit. We must average overall at least 25% gross profit on the selling price....

Every item we buy, old items included, should be price checked with other suppliers at least once a year on a routine basis.... Generally, do not switch unless new price is at least 5% better....

Do not keep a salesman waiting for more than five minutes without attention. If you are busy with another salesman, excuse yourself

David Grainger, in the Army Air Force and stationed at Boca Raton, Florida, joined a 1946 deep-sea-fishing excursion in Miami. From left: E. O. Slavik; the captain; Wally Burke, the Miami territory salesman, with whom Bill Grainger was very close; Teena Reinier; Bob Reinier, the Miami branch manager; the first mate; and David Grainger.

and go to the switchboard, greet the newcomer, and ask him to wait or come back later. Do not keep a salesman waiting while you are doing work at your desk; this makes him think you are trying to be important.

Take salesmen to lunch if possible, but always pick up the check. Never fail on this; it has paid big dividends. Do not accept any favors from any salesmen.... Give salesmen all the time they want but don't drag it on. If his price is high, tell him so, but no more, and don't discourage him. We want salesmen to feel they can come in any time. Phony rules, such as "salesmen interviewed 9 to 12," are out in our Company....

With our 39 branches we are entitled to a better price than a local jobber.... After price is quoted, always ask what you can do to get a lower price.

Ask if anyone anywhere gets a lower price. Then get it.... Do not ask if there is a cash discount. Assume there is one.[11]

At around this time, the company began a program designed to improve branch efficiency. It had been determined that greater efficiency could be realized by customizing the design of branch buildings to Grainger's needs. The first customized branch was opened in Detroit, and it was so successful that it became the standard for Grainger's branch expansion. Even so, the Detroit branch

Grainger catered to Americans' zeal for light entertainment following the lean war years by offering such items as movie cameras and accessories, which it sold exclusively to dealers.

CHAPTER FIVE: SCARCITY AND BOOM

didn't open without its own share of problems. Grainger soon realized that future branches would need adequate off-street parking and truck-high loading docks.

The Quota-Makers Holiday

In the fall of 1949, Grainger invited branch managers and salesmen whose sales exceeded pre-determined quota levels to the Chicago headquarters for the first Quota-Makers Holiday—a Grainger tradition that would grow in size, prestige, and popularity for many years to come.

"It was an enormous incentive program," remembered David Grainger. "It wasn't so much the money you made; it was the prestige. It became a huge fraternity. Nobody could wait to see all these people, and they really got to know each other. The whole administrative group in Chicago and their spouses were included, and it was just a big fun time."[12]

One year, because of its size, the event was split among four cities. But, as David Grainger pointed out, "It was a dud. The people from New Orleans didn't get to see the guys from Milwaukee or Miami or Seattle, etc., etc., and that wasn't any good."[13] Thereafter, the company held gatherings in Chicago until finally the event grew too large for any one facility to handle, and it was discontinued.

Growth Formula

In 1949, America was entering a halcyon age when it would lead the world in economic growth. In other words, customers had the money to spend and goods were readily available, and for Grainger this would prove to be a formula for unprecedented growth, diversification, and prosperity.

The onset of the Korean War caused a short-lived scarcity of motors during the early 1950s, yet Grainger was able to offer a ready supply of fractional-horsepower motors, which it advertised in special "Dealer Bargains" catalogs.

CHAPTER SIX

STAYING AHEAD

1950–1959

WHILE ECONOMICALLY THE 1950s was a golden decade for America, it was also a time of fundamental change. It was an era of Cold War fever, atomic bomb shelters, Senator Joe McCarthy, and "reds under the beds." Most middle-class workers received steady wages and easy credit, paving the way for a boom in the purchase of new homes, oversized cars, and new, labor-saving appliances. By the mid-1950s, the wealthiest nation on earth was producing nearly half of the world's manufactured goods and owned three-quarters of the cars, over half of the telephones, and a third of all televisions and radios.[1]

The Korean War

Not surprisingly, Grainger was well positioned to capitalize on this period of unrivaled prosperity. But at the start of the 1950s, the company had to contend with yet more restrictions arising from an overseas conflict. Less than five years after the German and Japanese surrender that promised a generation of peace throughout the world, war was declared by South Korea as North Korean forces stormed across the 38th parallel and raced to overrun the capital city of Seoul. The large-scale invasion took the world by surprise on June 25, 1950, and the United States quickly blamed the Soviet Union for precipitating the flagrant incursion.

Secretary general of the United Nations Trygve Lie called an emergency session of the Security Council, which passed one of the strongest resolutions of its four-and-a-half-year history, sharply protesting the action and demanding the immediate withdrawal of North Korean forces. Within days, U.S. President Harry S. Truman authorized the use of American armed forces, in association with other United Nations–member peacekeeping forces, to repel the invasion.

Once again, Grainger was taking care of its workforce. In June 1950 the popular *Grapevine* column "The Bosses Say" reminded employees that reservists and draftees would still have their jobs when they returned from service, without loss of profit sharing trust (PST) equity. On leaving for the service, employees were paid their full share in the PST, while, for retirement, their time in the armed forces would count as time working for the company.

When it became clear that materials would be rationed due to the Korean War, Grainger began

---

In 1951, the *Grainger Grapevine* announced that Grainger would need to move its headquarters from 740 W. Adams Street in Chicago because the building was in the path of the Northwest Superhighway, later renamed the Kennedy Expressway.

stockpiling inventory, and by March 1, 1951, its reserve of goods was the largest in its history. Motors were still very scarce, yet the company was able to fill most orders on fractional-horsepower motors. At the same time, fans and fan parts were in reasonable supply, though it was increasingly difficult to secure bearings and certain other fan components.

Once the restrictions were lifted, vast quantities of "war surplus" goods became available in the 1950s. These items were all high-quality military specification (MIL SPEC) matériel. The government clearly had a storage problem and quickly put in place a very efficient announcement and sealed auction process. While Grainger normally offered product lines with an ongoing supply, various "surplus" items were acquired and offered to its general customers. The results were outstanding. The customers enjoyed excellent value. Grainger enjoyed the revenue. But best of all, Grainger had the opportunity to quickly and inexpensively market test various products. Some of the successful items were portable engine generators, toolboxes, speed reducer gear units, gas engines, and transformers.

Moving with the Times

In the spring of 1951, the city of Chicago informed Grainger that it must vacate its office and Central Distribution Center at 740 West Adams Street due to the proposed Northwest Superhighway (later renamed the Kennedy Expressway). During the demolition the building caught fire, and television coverage showed the Grainger name vividly. A number of Grainger executives received worried phone calls.

The company had been forewarned of this possibility several years earlier and purchased an 84,000-square-foot lot at Loomis Street and Van Buren. The land was acquired as an optional building site should Grainger be evicted before locating to other quarters. Grainger eventually sold the land after it worked out a deal in 1952 with the Lincoln National

The company moved into the Oakley Boulevard/Adams Street headquarters and distribution center in October 1952.

Life Insurance Company of Fort Wayne, Indiana, to lease a 103,000-square-foot building at 118 South Oakley Boulevard/2330 West Adams Street. The Oakley/Adams building provided improved and expanded space for the general offices, the Chicago distribution center, and the Manufacturing Division.

Despite government restrictions, the company was able to move ahead with plans for branches in Baltimore, Toledo, Los Angeles, New Orleans, and Albany while also working on real estate deals in Pittsburgh, Cincinnati, and Kansas City. Opened for business on July 1, 1951, the new, 5,250-square-foot Los Angeles branch was immediately sold at cost to the Lincoln National Life Insurance Company in return for a 50-year lease, providing Grainger with improved cash flow for the purchase of merchandise. In a similar arrangement, it was fairly common during these years for employees to buy branch buildings and lease them back to the company. This practice provided a source of capital for Grainger and a good investment for the employees.

Embracing Technology

The Tabulating department had a meager but basic beginning in 1950. Maintaining a record on Kardex cards of each branch's sales and an inventory of each stock number was a major clerical task at the Chicago office. On the basis of this record, the Chicago product managers created shipping orders to the branches. Branch managers did not have input about what they received. The manual sorting of thousands of paper invoices was tedious and very time consuming. Each invoice was sorted as many times as there were lines on the invoice. Local high school students did the sorting part-time. And time was of the essence as prompt branch replenishment was the goal.

To mechanize and speed up this process, a tab card was keypunched for each line on the invoice: date, branch number, quantity, and stock number. The cards were sorted on a 420-cards-per-minute machine by branch number and then by stock number. A tabulating machine printing at 300 lines per minute produced hard copy used to post to the Kardex records. The Tabulating department consisted of six keypunchers, a sorter, and a tabulator. A breakthrough came two years later when the company installed a machine called an interpreter to print on the card what the holes meant. The time-consuming tabulator run was eliminated, and posting was done directly from the cards. The cards were then available for sales and statistical reports.

The Best of Times

During a business trip in the summer of 1951, Bill Grainger concluded a deal with the Ferrario Company in La Spezia, Italy, to purchase motors that would be sold under the Liberty brand name. Sales had increased by 51 percent over the first six

---

GRAINGER'S SONG

AROUND 1950, A ST. LOUIS EMPLOYEE named Jack Lambert penned the lyrics to "Grainger's Song." Set to the tune of "Might Lak a Rose," the anthem comprised two stirring verses intended to inspire company allegiance and ever higher sales:

Grainger is my comp'ny, Grainger is my work.
  I want to stay forever, and not a day
to shirk.
  May we be together, another year to come.
  And most of all forever, until our job
is done.

  May we be in good cheer always, never
fear of care and woe.
  And all our days be happy days, and
full of sales and "go."
  May our comp'ny stick together, from
here to kingdom come.
  Yes, most of all forever, until our job
is done.

The idea was for all employees to start their day with a lusty rendition of "Grainger's Song," but the anthem never caught on.

months of 1950, but a bill being considered by Congress would tax profits by 70 percent instead of 50 percent in order to help rearmament for the war in Korea. Some market watchers were predicting price rises and a return to the black market. Others were predicting surplus goods and falling prices. For its part, Grainger simply pushed onward and upward, courtesy of a commonsense policy of moderation: buying what it was selling, maintaining a normal inventory, and resisting a strong urge to speculate. In addition, Grainger had a very effective means of sales promotion that year, mailing dealers 1.5 million *MotorBooks* and 500,000 flyers—a marketing effort that cost $250,000.

Despite the Korean War, which was winding down by 1952, business conditions in the United States had remained fairly stable. For Grainger, stable conditions meant an average of 10,000 individual customer sales per week and 800 dealers going into the branches and making purchases each day. As a result, the company's contribution to the profit sharing trust in 1952 reached a record of more than $204,000.

Grainger, like many individual Americans in the booming 1950s, had more money than ever to spend on leisure activities. The company was quick to share its good fortune with its employees by encouraging branches to organize their own versions of the traditional Chicago-office summer picnic, to which all employees and their immediate families were invited. Grainger would pay for the site, evening meals, and prizes for children.

By this time, David Grainger had begun working full-time for the company after spending two years as an engineer at Franklin Electric, a motor manufacturer in Bluffton, Indiana. "My first day on the job, I came in without really having thought about what I was going to do," he recalled nearly half a century later. "I was 24, and youth is kind of dumb. Well, my dad laid the current *MotorBook* on the desk and said, 'You're a hotshot new engineer. Here's the catalog. Make it better.' So that's what I did, and I'm still doing it."[2]

On January 1, 1953, the New York branch opened at 533 Canal Street, near the entrance to the Holland Tunnel, and signaled the start of Grainger's regional warehouse program. The Canal Street premises provided Grainger with four more floors and 18,000 square feet of additional space to use as a regional warehouse to service the company's eastern branches. Around 400 items were initially stocked there for distribution to 10 eastern branches, while all other items remained warehoused in Chicago.

In April 1953, New York's sales volumes outstripped Chicago's, marking the first time that any branch had surpassed the main office. But in July, when 40 of the 46 branches exceeded their quota, Chicago gained the top spot once again, followed by New York, Los Angeles, Kansas City, and Dallas. The year 1953 was an extraordinary year all around, for more than half of the company's 46 branches surpassed their quota levels during the first two months, and overall sales figures were way ahead of those for the previous year.

Honing Sales Skills

Of course Grainger's employees, as well as the company's policies and the booming economy, were responsible for the company's stellar success, and Grainger was more than willing to help its workers excel and quick to reward them when they did. Merchandising Manager Russel "Rud" Francis was a case in point. Renowned as an excellent salesman, Francis wrote a column in the *Grainger Grapevine* describing how to identify large-volume motor buyers. Francis had joined Grainger in 1942 as a salesman at the Milwaukee branch and was moved to the Chicago office after achieving an enormous increase in Milwaukee's sales volume.

In February 1953, Francis was promoted to vice president of merchandising, and Ed Schmidt was named executive vice president.

David Grainger began working full-time for Grainger in October 1952. His father handed him a copy of the current *MotorBook* and said, "Make it better."

CHAPTER SIX: STAYING AHEAD

Right: Two years after her retirement in 1953 as company secretary, Margaret Grainger (left) posed with sister Laura Grainger. In 1933, Margaret had opened and managed the company's first branch, in Philadelphia.

Below: Rud Francis worked his way up through the ranks, starting as a salesman in 1942 and becoming vice president of merchandising by 1953.

The company's other corporate officers included Bill Grainger, president and treasurer, and his sister Margaret, who was the company secretary. Margaret would retire from the company just a few months later, after 26 years of service.

The company had five sales managers in 1953. Ray Ferguson's territory covered all branches east of the Mississippi except for New York and New England, which Herb Elfstrom covered. Walt Booth, based in San Francisco, was responsible for the Pacific Northwest. Phil Van Bussum's territory stretched from northwest of Chicago to the Rockies. And Howard Alder, promoted to Chicago that month from his job as a salesman in Salt Lake City, became sales manager for the region west of the Mississippi and south of Omaha.

At the start of 1954, a bulletin was issued to all branch managers and salesmen announcing sales quotas for the year and supplying instructions on how to reach them. For instance, sales representatives were informed that most of them had "spent too many calls on small 'peanut' customers or prospects.... You have to fish in bigger waters if you want to catch larger fish."

The branch managers were expected to be "sales-minded" and to "sell more than the customer came in to ask for.... Be pleasant over the phone. Treat all customers and prospects extra nice—they can go a lot of other places, you know, with their business."

With numerous branches conducting more business over the phone than over the counter, the company determined that its sales staff should brush up on the finer points of telephone etiquette. To that end, the November 1953 issue of the *Grapevine* featured a full-page article dispensing instructions on the subject.

The phone should be answered with the words, "GRAIN-GER, Smith speaking," with about an equal accent on each syllable of the word "Grainger," and in a slight singing tone. Do not add the words Company, Corp., Corporation or W. W. You may add, "good morning" or "good afternoon," if you like. Then give the caller plenty of time, and be thoroughly unhurried about the way you take the call....

After you have taken the order, suggest other related items in a tactful way whenever possible. People like reasonably aggressive salesmen. Those they don't like are the wishy-washy ones who sound as if they're afraid to talk....

Finally, the 1954 bulletin concluded by asking everyone to remember that they were involved in "the greatest and most fortunate industry—the electric industry, the future of which is unbounded, and almost unscratched. The growth ahead is almost unlimited for those who realize it and take advantage of it."

Bigger and Better

In September 1954 the Manufacturing Division moved out of the Oakley building and into its own premises at 3255 West 30th Street to provide needed space for Chicago operations. Encompassing three floors of 25,000 square feet each, the building was purchased for $240,000 and was owned by the PST, which leased it to the company. This brought the trust's real estate holdings up to a full one-third of its assets—the maximum allowed—while the other two-thirds was in liquid assets, primarily government bonds.

Grainger's fan sales represented a year-round business. In addition to being vital in the summer, fans were also in heavy demand during winter to ventilate barns, dairy stables, poultry houses, and other farm buildings. Fan manufacturers, farmers' associations, farming publications, and even the U.S. Department of Agriculture had spent a great deal of effort educating farmers on the merits of ventilating their outbuildings to reduce moisture.

When cattle and poultry were kept indoors during the winter, their body heat would cause moisture, which could eventually rot the building and rust equipment. Furthermore, excessive temperatures were a health hazard to livestock and could result in a significant decrease in milk and egg production.

The Dayton brand capitalized on this new market, and eventually some fans were designed specifically for farms and poultry houses. By 1955, Dayton was shipping record quantities of fans with each successive month, and fans were among the most heavily promoted items in the *MotorBook*.

Opposite: By establishing the branch and regional distribution center at 533 Canal Street in lower Manhattan, New York, Grainger was able to better serve its eastern customers while saving money on shipping.

Below: With strong sales growth, the Manufacturing Division moved in 1954 into its own building in Chicago, where it had much-needed additional space to manufacture all types of fans, fan parts, compressors, and pumps.

Branch managers and territory sales personnel whose sales exceeded predetermined quotas were invited to the highly anticipated, all-expenses-paid Quota-Makers Holiday, held each year in Chicago. The number of qualified people was small when the holiday began in 1949, but the group soon got so large that this very successful event had to be discontinued. The branch managers above were 1951 awardees.

Throughout these years, Grainger continued to set sales records on a daily, weekly, monthly, and annual basis. As it grew, the company had an ever increasing need for facilities to handle the storage and shipping of goods. In November 1956, the regional warehouse shipping program commenced operations, and during its first four months total billing amounted to $900,000, averaging 150 orders per week. In April 1959, plans were announced to stock the eastern regional warehouse—located at 533 Canal Street in New York—with practically every item the company handled, eliminating the 14 eastern branches' dependence on Chicago for inventory. The initiative also took advantage of overnight delivery from New York while reducing freight costs.

In September 1958, a contract was signed to erect a building for a new branch in Syracuse, New York. Syracuse was Grainger's 63rd branch.

In 1959 the merchandise service department (later renamed the customer service department) was established in Chicago. Managed by Frank Hoch, it handled repair parts for *MotorBook* items, particularly those bearing the Dayton brand name.

Reaping Rewards

The year 1959 was the first in which any employee—with the exception of Bill Grainger—reached 25 years of service for the company. Accordingly, it was announced that, thereafter, such individuals would be entitled to four weeks of vacation time each year.

By this time the Quota-Makers Holiday was an established annual event, increasing in popularity with each passing year. In 1957, for example, three simultaneous expenses-paid "parties" running from September 12 through September 14 were arranged at the Waldorf-Astoria Hotel in New York, the Mark Hopkins in San Francisco, and the Roosevelt in New Orleans. In a noble attempt at understatement, a sales department bulletin described these events as being "tops in glamour, reputation, and so well known that descriptive words are superfluous."

Meanwhile, in order to qualify for the "big fun party," all sales representatives and branch managers were urged to "oversell your quota every month—especially in January and February. Once you get behind, it's hard to get caught up. The rewards are many and plenty.... Go to work every day determined to beat your quota. You can't miss if your heart and soul are in it."

The Quota-Makers Holiday was characteristic of a company that continually came up with programs and policies to reward its workforce. The most visible of these was still the Profit Sharing Plan, and its continued popularity was partially due to the frequent amendments that updated it. The original policy, for example, stated that employees would start receiving their share at a specific year of age, even though they might still be working for the company and might prefer to wait until they retired. Beginning in mid-1955, however, the payout could be postponed at the discretion of the workers, and they would therefore not be taxed on the sum until they chose to withdraw it.

In September 1959, the *Grapevine* launched a column to discuss the fine points of the PST. The column pointed out that a pension was usually equal to a certain percentage of an employee's average earnings over a stated period of years. If an employee left a company before retirement, he or she wouldn't receive anything at that point, and the widow or widower of an employee often received nothing. The PST, on the other hand, didn't provide a specific amount at a specific time. Instead the company set aside a portion of its profits each year. The employee earned a vested interest and after 10 years would be entitled to the entire share. What's more, because the PST was set aside in a separate fund that wasn't available to the company, it was similar to a funded pension. (Nonfunded pensions depend on a company's willingness and ability to pay out at the appropriate time.)

In 1959, Grainger was able to contribute a record $373,697 to its profit sharing trust, $58,000 more than it had paid the previous year. This brought the company's total contribution to nearly $3.5 million since the plan's inception in 1941. Furthermore, in October 1959 the maximum benefits payable under Grainger's hospitalization, surgical, and medical plans were increased by a substantial 50 percent.

Clearly, the company was dedicated to the welfare of its employees, and though the following decade would be one of social, cultural, and political upheaval, the good times were about to get even better for those involved with W. W. Grainger, Inc.

Until the late 1990s, Grainger's branches were similar in appearance to this brownstone Manhattan branch on the corner of Washington and Canal.

CHAPTER SEVEN
GRAINGER GOES PUBLIC
1960–1969

THE EARLY 1960S WAS A time of advancement, affluence, and hope for many Americans. At the start of the decade, the income of the average citizen was more than a third higher than it was just 15 years earlier. The majority of Americans owned their homes, and nearly every family had at least one car. This was the "affluent society," and with improved economic mobility came a rapid exodus of families and industries from city centers to the suburbs.[1]

A Desire to Serve

As an aging President Eisenhower relinquished the Oval Office to the more youthful JFK, a social and cultural revolution was brewing in America. Moral attitudes were changing, long-held values were being questioned, and the old order was being challenged. For many, Grainger represented a comforting oasis of conservative stability.

"One of the things that I remember about Grainger when I first joined it is how much it was like a close-knit family organization," said Dale Woods, who began working at the Des Moines, Iowa, branch in 1959 and later became a regional sales manager. "At Christmastime you always got a signed Christmas card from Mr. W. W. Grainger, which I thought was a nice touch, and you also got a ham shipped to you with your freight. That made a lasting impression. Even though I started in early December, I got a ham the first year, so I was impressed."[2]

Rich Greenlee, regional operations manager, remembered when Bill Grainger visited the Denver branch, where Greenlee first worked. "He'd come in and talk to all the employees. It was really a morale booster. You had a sense of real ownership because you felt you were a part of it all."[3]

The company in fact had a reputation for engendering a family spirit, and Bill and David Grainger were often described by employees as paternalistic. "I remember walking into the branch that was connected to the headquarters and seeing Mr. [Bill] Grainger lying on the floor showing one of the branch managers who had just come back from back surgery how to do exercise," said Angie Salazar, who began working for Grainger in 1952. "We didn't have air conditioning back then, and he used to send out for ice cream on break for everybody."[4]

Grainger's first public offering of stock, in March 1967, met with great enthusiasm and success, as this news item in the *Wall Street Journal* attests.

Dick Quast, who began his Grainger career in 1957 and later became vice president for real estate, also remembered how personable Bill Grainger was. "Bill would go out in the warehouse and shake everybody's hand, and everybody knew him," Quast said. "He didn't have any pretentiousness about being the owner of the company. He was a down-to-earth guy, and everybody liked him. Sometimes he would slip them some money, saying, 'Here, take your wife out to dinner.'"[5]

The feeling that employees were respected and well cared for contributed greatly to employee loyalty, as did the company's unswerving code of ethics. "Here was a company that supported the straight-arrow approach," said Max Mielecki, who retired as vice president of advertising in 1996 after 32 years with the company. "There was an honesty and a lack of politics that was very refreshing, very encouraging."[6]

Mike Kight, a Grainger vice president who began working in the Albuquerque branch in 1967, remembered that Bill Grainger stopped by the branch to pick up some lightbulbs for his hotel room so he could see better. "It was a very small sale," said Kight, "so we told him we'd just write it off for branch use. But he said, 'No, I'm going to pay you for them. I'll give you cash.' Small as it might have been, he was going to pay for the purchase. He had such a high level of integrity, and that's kind of what drove the culture."[7]

"What made working for Grainger a very positive thing was the attitude," added Dale Woods. "The Grainger people always had that can-do attitude, and you'd see it in the field, you'd see it in Chicago. When I was hired, one of the first pages in the hardbound price catalog said that to be in the Grainger employ you had to have a burning desire to serve, and that always stayed with me."[8]

That burning desire to serve was something that Grainger helped inspire by rewarding employees with ancillary benefits that went beyond what most Americans might expect from their jobs. In addition to the profit sharing trust, generous health insurance, and myriad other incentives and rewards, Grainger established a college scholarship program in October 1961 for the children of Grainger employees. Sons and daughters of deceased, disabled, or retired employees were also eligible, but children of the Grainger family and company officers were excluded. Sponsored by The Grainger Charitable Trust, which had been established by Bill and his wife, Hally, the program provided scholarships toward tuition at any fully accredited university or college in the United States. Full-time attendance was a requirement, any degree course could be pursued, and the cash award—ranging from $150 to $1,500 annually, depending on the employee's income—could cover tuition, books, room, or board. Grainger allocated its first scholarship for the school year commencing September 1962.

The company introduced another incentive in January 1966—the team incentive program (TIP), whereby employees were eligible to win $117.50 in cash awards for the seven months from January through July, should they meet their sales quotas. The idea of TIP was to help each employee appreciate the importance of every customer order as well as the value of good service on the part of branch personnel. In 1967, the program was broadened to include all 12 months, while individual incentives were increased to $300.

In November 1966, the company revamped its hospital and medical health plan. The new plan provided expanded protection to all employees at the company's expense. In addition, major medical coverage (also known as catastrophe coverage) was included, kicking in where ordinary coverage left off and providing up to $9,500 for extended treatment.

Leaps and Bounds

The rewards and incentives Grainger was able to pass on to employees reflected more than just satisfaction in worker performance. The company was, in fact, performing better than ever. In 1960, Grainger's sales reflected the largest dollar increase in recent history.

Part of this success was due to pump sales, which had risen steadily over the years and went through the roof in March and April 1960, when much of the United States was inundated by floods. The General Pump Company produced pumps exclusively for Grainger, and Grainger in turn supplied General Pump with many of the components, including the motors. During the three-week period ending April 13, General Pump shipped Grainger six complete trailer loads, which totaled more than 3,650 pumps. It was the heaviest production

In 1962, Grainger relocated much of its operations, including a new branch, to 5959 West Howard Street in Niles, Illinois. The company's downtown Chicago branch remained open at 2330 Adams Street (above).

run since the Indianapolis-based operation had been established.

As sales volumes grew, so did Grainger's need for more space. In June 1961, Grainger paid $600,000, plus a vacant 33-acre tract of land on Mannheim Road that it had owned since 1958, to acquire a 17-acre lot with a 130,000-square-foot, one-story building at 5959 West Howard Street in Niles, Illinois, a Chicago suburb. Purchased to house Grainger's general offices and Central Distribution Center, the land provided needed room for expansion. The September Quota-Makers Holiday included a tour of the new building, and by November 19, 1962, Grainger's Central Distribution Center, customer service department, and data processing department were in residence.

"Before the move to Niles we never had enough room," recalled David Grainger. "We never had enough docks, and we couldn't even ship a whole truckload. We shipped four or five times a day to the same branch. We spent a ton of money on shipping, but when we finally moved to Niles, we were able to take 2 percent out of our costs."[9]

The company's branch operations expanded as well. Though in later years branch locations would be determined through extensive study, early expansion was largely unrestrained since every branch opened proved successful if well managed. Dick Quast remembered how locations were selected. "In those days, when we needed a branch, [Bill Grainger] would call the sales managers and say, 'Okay, each one of you guys give me a list of five locations where we should have branches next year.' And they'd give a list. The thinking was that if we were there, they would come."[10]

In 1963, Grainger began its most ambitious construction program to date, with 11 new branches planned for the year. Two of these were built in entirely new market areas, while nine buildings were erected in existing branch markets. Another seven branches received additions. In all, Grainger had 77 branches by the end of 1963 and 802 employees nationwide. That year, 42 branches exceeded their quota figures, 30 were under (although 24 of them missed the mark by less than 5 percent), and the company's total assets reached more than $18.5 million. By September 1965, Grainger had no fewer than 34 building projects planned around the nation.

Grainger set many new sales records, including the summer 1963 high of $39,065 for the week of June 10–15 and a record $48,157 during the December 16–21 countdown to Christmas. Then in the four-week period from November 30 to

December 26, 1964, Grainger handled a record total of 174,752 individual sales invoices, averaging 9,197 invoices per day, 1,150 per hour, and 19 per minute. During this time, a Grainger sales invoice was being completed somewhere in America every three seconds.

Meanwhile, the company continued its rapid expansion. When Grainger first purchased the 130,000-square-foot building in Niles, 17 acres had seemed a vast expanse in which to grow. But by the summer of 1966, Grainger had grown so much that it had to add another 5,100 square feet of office space, and the warehouse expanded to 310,874 square feet, leaving only eight acres to build on.

Grainger was also constructing a 50,000-square-foot single-story building in Cranford, New Jersey, to serve as the eastern regional customer service distribution center. Merchandise was transferred there from the upper floors of the Canal Street building in Manhattan, the multistory, crowded, elevator setup that had previously served the company's eastern branches and customers. The new structure's facilities were far superior, boasting automatic loading-dock levelers that provided high-dock loading and unloading for any truck-bed height.

Always the Best

As it grew, Grainger continued to keep pace with technology. In April 1961, the company installed its first wide-area telephone service (WATS). WATS saved Grainger time and money by charging a flat hourly rate for all calls made to numbers outside Illinois.

The company also upgraded its computers. In 1961, the *Grainger Grapevine* ran a photo feature about the Univac Solid State STEP computer being installed in the Tabulating department at Oakley Boulevard. STEP was an abbreviation for "simple transition to electronic data processor," and the Solid State Univac replaced the department's slower, vacuum-tube models.

According to the article, the new computer was "really four machines in one! It will read tab cards, calculate, print reports, and punch tab cards, and do all four operations simultaneously. Units we now use calculate information in thousandths of a second, but the STEP computer prints six times faster, too. Our present printer runs off reports at a rate of one hundred lines per minute.... Soon these same reports will be run off at six hundred lines per minute."

"It was a big honker of a thing. It could eat half the block," remembered Edward Bender, who was one of Grainger's first "internally programmed computer programmers" and was in charge of operating the Univac. "The first application that I remember on it was the accounts receivable. It used to take days to process these things, but with an electric computer, we were able to do it in a quarter of the time."[11]

By the mid-1960s, Grainger had upgraded its computer system from Univac to the state-of-the-art IBM 360, and though Ed Bender said that converting between the two systems was "the nastiest job I ever had at Grainger," the switch, in his opinion, was "a major decision in the company's history.

With the number of branches, products, and sales invoices increasing every year, employees in Grainger's record-keeping department seldom rested.

If we hadn't made the switch, it would have probably set us back big-time at a time when we were growing at 15 to 20 percent a year."[12]

Grainger remained at the forefront of technology. By 1966, the company transitioned to the IBM 360, a significant space saver compared to the Univac units it replaced.

Golden Dolphin

While Grainger was not actively seeking new areas of growth beyond branch expansion and additional catalog items, David Grainger became involved with a product line other than electric motors. It was losing money, but David thought it could become profitable through Grainger's distribution expertise. That's how an enterprise known as Golden Dolphin came on board.

The Golden Dolphin program dealt in a line of fashionable, color-coordinated bathroom accessories such as towels, rugs, shower curtains, and, later on, toilet seats, toilet tissue holders, toilet brushes, soap dishes, mirrors, and assorted porcelain accessories. Sales of these items were made to department stores and specialty shops across the country, with shipments originating from the Central Distribution Center in Niles. In April 1967, the Golden Dolphin trade name became a wholly owned subsidiary of Grainger called Coordinated Sales, with salesmen selling directly to department stores and specialty shops rather than through Grainger outlets.

Grainger ran Golden Dolphin for more than a decade. But after years of continuing losses, partially because a handful of the big retail stores it sold to weren't paying their bills, Golden Dolphin would halt operations in December 1978. "We finally just said 'sell it,'" said David Grainger, who acknowledged that the entire operation was a learning experience for him. "I was a naive guy in the industrial supply business who found out what the retail industry was like."[13]

Prudent Operations

Golden Dolphin may have been vulnerable, but Grainger's core operations continued to be very successful. In 1965, sales volume exceeded that of previous years for the 22nd consecutive year. There had been only three years—1938, 1943, and 1944—when Grainger, because of material shortages, had not exceeded its revenues for the previous year.

During July and August 1965, David Grainger and Ed Schmidt exchanged several letters regarding the value of entertaining and maintaining close friendships with suppliers. Statistics were readily available for Grainger's top 10 annual suppliers. During the 21-year period from 1944 to 1964, a total of 41 different suppliers had made the list, and in his correspondence to David Grainger, Schmidt concluded that there was little correlation between providing entertainment, such as golfing and dining out, and continuity of suppliers. More important, Schmidt decided, was the mutually beneficial relationship between buyer and seller.

Grainger featured its Golden Dolphin line of coordinated bath fashions in department stores and specialty shops and even set up displays at Chicago O'Hare International Airport.

Schmidt cited several examples of big clients who had been wined and dined at great expense over the years, only to drastically reduce orders for reasons of their own financial gain. Nevertheless, he noted that "much less entertaining is done today for our 'Top 10' than was the case 15 to 20 years ago. Yet I would say that our position with nine of the 'Top 10' ... is much more solid than was the case 15 to 20 years ago."[14]

But it was just that sort of difference in management style that kept Grainger on its toes. "There was an interesting relationship between W. W. Grainger and E. F. Schmidt," said Max Mielecki. "W. W. was Mr. Outside and E. F. was Mr. Inside. Every time Bill Grainger got too far out, Ed would pull him back in."[15]

Grainger Goes Public

W. W. Grainger, Inc., was quickly evolving. By December 1966 it seemed that a new epoch was fast approaching as management began considering a bold new step.

As the Vietnam War raged overseas, the late 1960s was a time of social and cultural upheaval in America. Chicago would witness its own insurrection in the form of riots at the 1968 Democratic National Convention. Even so, the company officers were preoccupied by other matters, for after years of discussions the company was finally going public.

The idea of a public offering had been floated as early as May 1961 at a meeting among David Grainger, Bill Grainger, Ed Schmidt, and Jack Nicholson, of the accounting firm Alexander Grant & Company.

"Throughout the early 1960s, my dad would take the train to work from Hinsdale a good deal of the time, and he'd continually meet these people who'd want him to sell the company," David Grainger recalled. "He'd get to the altar, but then he'd always back off, saying, 'I can't trust these people to handle the suppliers, the customers, or the employees properly.'

"As for going public, my mother thought it was going into the land of the unknown, and I think she was a great influence."[16] Meanwhile, Maury Stans, who then owned Alexander Grant & Company, had also been talking to Bill Grainger about the benefits of going public. Both men lived in Hinsdale, and they had been close friends since the 1930s. On occasion, Stans would come to the Grainger home for dinner, and "just like somebody looking for money for charity, Maury would always bring it up," recalled David Grainger.[17]

"You ought to take the company public," Stans would inevitably say.

After repeated dinners that all came back to the same conversation, Bill Grainger finally said, "Maury, you need to give me a good reason to go public."

"Well, it would be great experience for David," Stans said—and as David Grainger pointed out, Stans was correct.[18] David Grainger stated in a note to Ed Schmidt that the principal motivation for the sale would be to diversify the family's holdings.

But Bill Grainger was still not convinced. "I have always [been] and still am against going public unless it definitely is most beneficial to the company," he asserted in a handwritten note to his fellow directors on April 26, 1966.[19]

The Slaviks, for their part, also had to be convinced. Stans, who represented the investment firm of Glore Forgan, Wm. R. Staats, Inc., negotiated between the Slaviks and the company regarding details of the offering.

"My father was very strong willed like my grandfather, and I know there were times when there were some contentious meetings, but he always looked at Grainger as a long-term investment," E. R. Slavik's son James recalled.

I remember him talking to us as a family about Grainger going public, and he was very excited about that. He instilled the feeling in me and my brothers that this has been a good company for us and that we should remain involved with it. He was very proud of Grainger and proud of being involved during the public offering. He was also very proud of Grainger's board. He really respected them, and he derived a lot from knowing them as well as from contributing himself.[20]

By late June 1966, the Slaviks had agreed in principle to sell 30 percent of their holdings as long as all other parties did the same. On November 9, 1966, David Grainger, E. R. Slavik, Ed Schmidt, and Maury Stans signed a preliminary underwriting agreement for a public offering of capital stock in W. W. Grainger, Inc.

As a public company, Grainger has been traded on the Chicago (formerly Midwest) Stock Exchange, the American Stock Exchange, and the New York Stock Exchange. This certificate was issued for 100 shares on September 17, 1975, on the New York Stock Exchange.

On February 17, 1967, Grainger filed with the Securities and Exchange Commission in Washington to sell 720,000 shares of Grainger common stock to the public. Following the sale, the selling stockholders would own none of the company's common stock but all of its Class B stock, and none of the sale proceeds would accrue to the company.

The company's officers knew that such a profound event was bound to raise concerns of employees, so a March 1 operating bulletin reassured them that they should not feel threatened by the impending sale and needn't purchase stock: "An employee's standing with the company will not be affected one way or the other by buying or not buying stock," the bulletin stated.[21]

On March 29, 1967, just six months shy of its 40th anniversary, W. W. Grainger, Inc., officially became a public company. A nationwide syndicate of 132 investment bankers headed by Glore Forgan, Wm. R. Staats offered the 720,000 shares at $19 per share, and within a few hours all of them had sold.

At the time, Jere Fluno, who would join Grainger in 1969 as controller, was working for Alexander Grant & Company and was involved in taking Grainger public from an accounting standpoint. "Traditionally it had been such a private company," he recalled. "Maury Stans had sold [Grainger] on going public, and initially I don't think they understood all of the implications. The problem was that there were a number of people sitting on a tremendous amount of wealth, and yet it wasn't marketable."[22]

Grainger's new board of directors included Maury Stans and John E. Jones in addition to Bill and David Grainger, Ed Schmidt, E. R. Slavik, and

Graydon Ellis. Maury Stans shortly left the board to become Secretary of Commerce under President Richard Nixon. Kingman Douglass, of Glore Forgan, Wm. R. Staats, took Stans's place. Douglass had been heavily involved with the public offering and would be a director for 25 years.

In August 1967, the board elected four new officers: Herbert O. Elfstrom, vice president, advertising; Robert H. Lollar, vice president, branch operations; Edwin C. Zimmer, vice president, treasurer, and assistant secretary; and Lee J. Flory, vice president, corporate secretary, and assistant treasurer. Grainger's other four officers were Bill Grainger, president and CEO; Ed Schmidt, executive vice president; David Grainger, vice president; and Rud Francis, vice president, merchandising.

Nine months later, at the December meeting of the board, Bill Grainger, who had founded the company as a mail order business in the year of his son's birth, tendered his resignation effective January 2, 1968. David Grainger noted that his father's retirement had nothing to do with going public, despite the timing.

"At that point in his life, Bill was over 70, and one of the things I think he wanted to do was just get away from the day-to-day running of the company," said Lee Flory.[23]

At the same meeting, David Grainger was elected to the newly created post of chairman of the board of directors, and Ed Schmidt was elected president. Bill Grainger would continue to serve as a board member.

Flush with Success

The company's 1967 annual report and proxy statement, its first as a public company, was distributed to approximately 3,500 shareholders. During the year, sales rose to $88.6 million from $80.2 million the year before, while net income rose to $3.92 million from $3.89 million.[24] Ten-year comparative figures in the report illustrated that from 1958 to 1967, sales had increased 322 percent, net profits had shot up 570 percent, the book value per share had grown 345 percent, working capital had gone up 355 percent, and the number of branches had increased 50 percent. During the same period, each year's profits had topped those of the preceding year. In 1968, sales exceeded the

In 1968, David Grainger, the company's new chairman, and Ed Schmidt, president, discuss features of an electric motor distributed by W. W. Grainger, Inc.

Grainger's Central Distribution Center in Niles was a model of modern efficiency. The first two levels were used for stock picking, while the upper two levels were for reserve stock.

$100 million mark for the first time, while earnings rose to $4.8 million.[25]

Not too surprisingly, what had always been a quiet and private company was beginning to attract widespread attention. The July 12, 1968, edition of the *Chicago Sun-Times* contained a story about Grainger in its financial section. Written by the paper's financial editor, the article provided an overview of Grainger's distribution network and a compendium of sales and earnings figures, along with a history of the company and its founder. Noting that from 1958 to 1967 the company's average annual compounded growth rate for earnings had been 15.8 percent, the profile concluded that Grainger "has one special growth factor going for it in addition to population and economic growth."[26]

After the public offering, the company set out to aggressively grow the business. In late 1968, the Central Distribution Center in Niles expanded by 75,000 square feet, then by 76,000 square feet in 1969, bringing the total to 440,000 square feet. Also in 1968, the company opened four new branches, a 34,700-square-foot distribution center in Fort Worth, Texas, and a 34,600-square-foot distribution center in Oakland, California.[27] In June 1969, Grainger's 100th branch opened, in North Milwaukee, and by the end of the year, the total number of branches had reached 104.[28]

By the end of the decade, Grainger was selling more than 4,500 different items to approximately 300,000 customers, with motors accounting for about half of sales. Blowers, fans, pumps, and air compressors accounted for another 25 percent. Because many of these items contained motors as an integral part of their construction, Grainger's motor sales grew more strongly than ever.

Accordingly, in July 1969 Grainger expanded into the manufacture of electric motors and grinders through its acquisition of Doerr Electric Corporation of Cedarburg, Wisconsin, and three affiliated corporations, two in Sturgeon Bay, Wisconsin, and the other in Burr Oak, Michigan. "We were taking some two-thirds of Doerr's output, and it was a natural fit for both companies," explained Ed Schmidt a year after the deal was finalized.[29] As part of the deal, Doerr's president and cofounder, Lee Doerr, took a seat on the Grainger board of directors while continuing to run Doerr as a distinct Grainger operation.

Within a few years Grainger would divest itself of its manufacturing facilities in order to concentrate on its core competency—distribution.

As part of the 50th anniversary celebration in 1977, Bill Grainger was honored with a collage of pages from the *MotorBook* that illustrated how the products had changed as the company evolved.

CHAPTER EIGHT

EXPANSION AND REALIGNMENT

1970–1979

THE 1970S WERE DIFFICULT for the American economy. The beginnings of the Watergate scandal, the growing conflict in Vietnam, and the often violent reaction to the Civil Rights movement had already soured the country's mood. The energy crisis of late 1973 caused severe price inflation and forced conservation to the top of the national agenda. Unable to withstand the pressure, the U.S. economy plunged into a recession, suffering a unique combination of rising inflation and recession.[1] From 1973 to 1974, consumer prices rose 12.2 percent and continued to rise until 1978.[2] As new technology replaced large sections of the workforce, unemployment took a dramatic leap. Price increases overtook wage raises for millions, and a report by the U.S. Census Bureau estimated that about one-tenth of Americans lived below the poverty level. This decline in living standards in turn contributed to a soaring crime rate.[3]

While many businesses remained stagnant or slid downhill in the floundering economy, Grainger would experience few cutbacks and would, in fact, continue to expand, diversify, and restructure. In July 1975, Grainger's common stock made its trading debut on the New York Stock Exchange, providing Grainger with more prestige and wider media coverage. (Previously the company had traded on the Midwest Stock Exchange, a Chicago exchange reported mainly in midwestern newspapers and financial publications.)

Elbow Room

Grainger's growing network of local branches and full-time territory sales staff enabled it to serve broader markets while still maintaining prompt, efficient, and economically viable individualized service. Local branch managers, for example, knew when factors such as weather or prevailing business conditions might affect inventory and could control the quantity of each item stocked in their branches. Likewise, the salespeople in the field maintained close relations with customers to determine their individual needs, communicating with branches and the corporate office to make sure Grainger could provide the products in demand.

While continuing to strengthen customer relations, Grainger's operations were rapidly growing. During 1971 alone the company opened 14 new branches, bringing its total to 122. By that time Grainger offered more than 5,500 different

Grainger celebrated its first half century of unprecedented growth with a special Quota-Makers banquet in Chicago.

products to 350,000 customers—and those numbers were growing every year.

In 1971 Grainger acquired an 11-acre site that adjoined the Central Distribution Center in Niles to meet future expansion needs.[4] In September 1973, ground was broken on the site for a 230,000-square-foot addition to the Central Distribution Center. This was Grainger's largest addition yet, and it brought the Chicago headquarters and Central Distribution Center to 690,000 square feet.[5]

Still the company needed more space to keep pace with its rapid growth. In June 1974, Grainger purchased a 470,000-square-foot warehouse in Bensenville, Illinois, pushing the Chicago headquarters and warehouse floor space to more than 1,000,000 square feet.

Meanwhile, Grainger continued to expand its manufacturing arm. Throughout the early part of the decade, the Doerr plants in Cedarburg and Sturgeon Bay, Wisconsin, were repeatedly enlarged

CHAPTER EIGHT: EXPANSION AND REALIGNMENT

Right: Though Grainger would eventually discontinue manufacturing, in the 1970s the company still produced Speedaire and Dayton industrial products in its Northbrook, Illinois, plant.

Below: After Grainger purchased an additional warehouse in 1974, its Chicago-area distribution space totaled more than one million square feet. Often merchandise traveled more than a half mile from its picking location to the shipping docks.

to accommodate demand for motors. In 1972, the Manufacturing Division was relocated to a 180,000-square-foot facility in Northbrook, Illinois, where it had 40 percent more space to produce items including fans, compressors, and pumps.[6]

That year Grainger acquired McMillan Manufacturing Company, of St. Paul, Minnesota, which had been supplying Grainger with shaded pole motors for more than 25 years. McMillan's motors added to Grainger's already reputable lineup of trademarked goods, which included the brand names Dayton, Speedaire, Teel, Dem-Kote, Demco, and Doerr. In 1975, the McMillan operations were discontinued due to low profits, while Grainger's newly constructed Doerr plant in Anamosa, Iowa, began producing general purpose fractional-horsepower motors.[7]

Technological Know-How

In addition to modern materials-handling systems at its warehouses, Grainger installed state-of-the-art data-processing systems at the main Chicago office and the Doerr manufacturing headquarters in Cedarburg, Wisconsin. The data-processing systems enhanced inventory control, order processing, production scheduling, requirements planning, and accounts receivable. In 1976 Grainger installed programmable data-processing terminals, called Data Point computers, in all its branches and manufacturing plants

Grainger's advanced data-processing systems allowed it to keep pace with the growing business.

to connect to an advanced central computer in Chicago. The new computer network improved customer service by speeding up branch ordering and invoicing and improving inventory control.[8]

"We had developed probably the leading-edge system in the country because we had automatic application," said Robert Collins, retired vice president of branch operations. "We had everything in that system."[9] Indeed, during 1977, the branch terminal system was processing an average of more than 23,000 invoices each business day.[10]

George Rimnac joined Grainger in 1978 and was impressed with how technologically advanced the company was. "I think the person who was personally most responsible for that was David Grainger himself," said Rimnac, who was hired because he had experience in programming Data Point computers. "The company had a high degree of pioneering spirit in terms of the kinds of applications they tried on computers. The reason they were putting Data Point computers in the branches was actually a pretty revolutionary idea in 1978, but they knew that with so many branches, they had to have a communication network."[11] Rimnac went on to become a vice president and chief technologist at Grainger.

Legacy of Leaders

During these years of rapid growth, the company experienced some notable arrivals and departures. On February 18, 1973, Elmer O. Slavik passed away in California, just two months after his wife's death. E. O. was the father of then board member E. R. Slavik and grandfather of future board member James Slavik. Instrumental in Grainger's early development, he had moved to the West after retiring as Grainger's executive vice president in 1948. In the early 1970s, he had taken his son's place on the board for a short time when E. R. was experiencing heart problems.

"My grandfather was very family oriented and interested in passing wealth down to the next generations," Jim Slavik recalled. "He was very forward thinking about that, and he and my father did a lot of work to be tax efficient about how they planned as they looked for ways to help future generations in our family."[12]

In December 1973, President Ed Schmidt announced his mandatory age-65 retirement as company president. In January 1974 he passed the reins to David Grainger, who had been chairman of the board since his father's retirement in 1968. Schmidt was vice chairman for two years, until his death on March 30, 1976, at age 67.

"The legacy that W. W. left, that Ed Schmidt left, that E. O. Slavik left, was a firm foundation on which to build," Jere Fluno said in a 1998 interview. "They left a balance sheet that looked like Fort Knox—and it still does. Even today, what you see is what you get. There are no games played with the numbers. The accounting is very, very conservative, and that's largely thanks to those guys."[13]

Fifty Years of Success

W. W. Grainger, Inc., had come a long way since its humble beginnings inside a loft on 22nd Street in September 1927. Back then, the first *MotorBook* catalog had predicted a "Bright Future for Dealers Selling Motors," yet not even in his most far-flung dreams could Bill Grainger have envisioned the scale and durability of his venture's success.

In 1977, its 50th year, Grainger sales were $498.5 million, a 25 percent increase over the previous year. Earnings increased by 28.1 percent to $33.4 million. The company's 141 branches in 43 states and the District of Columbia were being supplied by more than 1,300 outside vendors and represented by 424 salespeople, with nearly 4,900 men and women on the Grainger payroll.[14]

Over the past five years, Grainger had been growing at a rate of about 20 percent annually and was fast approaching the milestone of a half billion dollars in revenue. Meanwhile the company had contributed $36.8 million to the profit sharing trust since its inception in 1941, paying out more than $13 million to retired Grainger employees.

In contrast to the initial eight-page 1927 edition of *MotorBook*, which boasted 41 items, the 1977 edition contained 796 pages and offered 7,600 products and accompanying technical data to the company's more than 570,000 customers.[15] Before adding any new product—whether as a result of product research or on a recommendation from an employee, customer, or vendor—Grainger's engineering department conducted numerous tests to ensure that it met the company's strict requirements.

The 1976 *MotorBook* featured approximately 1,400 models of electric motors, electric motor controls, gear motors, fans, blowers, liquid pumps, air compressors, air tools and paint-spraying equipment, power transmission components, heating equipment and controls, air conditioning and refrigeration equipment and components, gas-engine-driven power plants, portable and stationary power tools, mechanics' hand tools, office equipment, material handling and storage equipment, plant and office maintenance equipment, lighting fixtures, and lawn and garden equipment.

About 600 new items were added in 1977, enabling each branch to stock up to 7,500 items for immediate availability to distributors, dealers, contractors, service shops, industrial and commercial maintenance departments, and all customers within its sales territory. Furthermore, a company inventory of more than 25,000 replacement parts helped provide a reliable backup service.

W. W. Grainger, Inc., was indeed maturing. During a banquet at the Quota-Makers sales meeting in Chicago in September 1977, Bill and David Grainger made the first cut into a giant 50th-anniversary cake. They were treated to a standing ovation for the contributions they had made to the lives of employees, customers, suppliers, and the electrical industry as a whole.

Just a few months later, the younger Grainger was cutting another anniversary cake during the 36th Annual Service Recognition Luncheon at the Ritz-Carlton Hotel in downtown Chicago. The event was attended by 191 Chicago office employees with five or more years of service, along with 58 new salespeople. By then, 899 Grainger employees had completed five or more years of service, and the company had 46 active members in its Quarter-Century Club.

Even Bigger

In 1978, after expanding the Central Distribution Center by leasing a 213,000-square-foot building in Elk Grove, Illinois, Grainger began two significant construction programs at its Chicago

By the late 1970s, the annual Quota-Makers gathering had grown from 400 to 600 attendees. It eventually grew so large that Grainger had to discontinue it because no area facility could hold so many people.

headquarters. First, it remodeled and renovated the office building at 5500 West Howard Street in Skokie, Illinois—purchased a few years earlier—to add 65,000 square feet of office space. In July 1979, Grainger moved its corporate headquarters functions to the Skokie location. A few departments remained at 5959 West Howard in Niles, about half a mile west of the new complex.

The second program marked the company's entry into automated warehousing and distribution. Grainger was ahead of its time in terms of technological achievements. Finding that it needed more warehouse space in Niles but had little land to expand outward, Grainger decided to expand upward and began building a fully automated storage structure—an automatic storage and retrieval system (AS/RS)—at the south end of the Central Distribution Center. When completed, the facility's 84,000 square feet seemed small compared to Grainger's other warehouses, but it was 68 feet high and handled 32,000 pallets, as much inventory as a conventional 250,000-square-foot warehouse.

Grainger was undergoing some internal restructuring as well. On December 11, 1978, the board of directors approved Grainger's realignment into a divisional organizational structure, with a corporate office, Distribution Group, and Manufacturing Group. As part of the restructuring, Wiley Caldwell was appointed president of the Distribution Group, and George Mathews became president of the Manufacturing Group. Caldwell, who would become president in 1983, had joined Grainger in 1977 as vice president of operations. Other officers elected

In 1977, Bill and David Grainger cut an enormous anniversary cake celebrating the company's 50th year.

included David Grainger, chairman and president; Jere Fluno, vice president of finance and treasurer; Don Hansen, vice president of administration and planning; Lee Flory, vice president and secretary; Dick Quast, vice president of real estate; and Robert Pappano, controller.

The merchandising, advertising, and data processing departments were already key parts of Grainger. Merchandising, led by Richard Norman, was responsible for the selection, pricing, and promotion of products. The advertising department was run by Max Mielecki. It was responsible for the *MotorBook* and various internal publications in addition to advertising in trade publications.[16]

By the end of the decade, Grainger had 152 branches in 46 states, nearly 500 territory sales representatives, more than 8,500 products, and 650,000 active customers.[17] Yet the company's mammoth proportions would seem small in another 10 years, when the numbers would have more than doubled. Far from leveling out, Grainger's expansion program was picking up even more steam.

Throughout the 1980s, Grainger expanded its distribution presence by opening new branches, renovating old ones, and moving others to more strategic locations. In 1980 alone, Grainger opened 14 new branches and moved 15 others to more modern facilities.

CHAPTER NINE

MARKET PENETRATION

1980–1989

RONALD REAGAN'S RISE TO the presidency in 1981 brought far-reaching changes in the United States. Consumer spending soared throughout the decade, but financial woes struck millions of Americans following the collapse of hundreds of savings and loan institutions and the "Black Monday" stock market crash of October 14, 1987.[1]

But Grainger remained a model of business success. Throughout the 1980s, Grainger provided for future growth and productivity increases by expanding and modernizing its distribution branch network. In 1980, the company opened 14 new branches, while 15 branches relocated to larger, more modern facilities and 12 branches added floor space.[2] Over the years, Grainger continued to add more branches and space to its Distribution Group so that by the end of 1985, Grainger had 188 branches in 48 states and the District of Columbia, and its operations encompassed 3,922,000 square feet.[3] That number would seem small by the end of the decade.

The only downturn Grainger experienced during the 1980s came in 1982, when the company's traditional growth mechanisms began to slow. That, combined with the unstable economy, caused Grainger to experience its first decline in sales and earnings since 1943. Sales in 1982 were down 6.6 percent to $803.2 million, while earnings declined 10.9 percent to $50.1 million.[4]

Though sales and earnings rebounded the following year, David Grainger and company officers took the loss as a sign that they needed to think more strategically. During the decade they initiated the company's first long-term planning scheme, which led Grainger away from manufacturing and toward being a market-driven company.

Grainger continued achieving healthy growth rates throughout the rest of the 1980s by doing what it did best: helping customers minimize expenses by offering one-stop shopping for maintenance, repair, and operating supplies (MRO), or as David Grainger put it, "having the right thing in the right place at the right time."[5]

As Grainger broadened and intensified its ongoing expansion program, it also kept pace with technological developments and performed marketing studies that helped improve customer service, increase efficiency, and ensure future growth. At the

---

In 1985, Grainger's catalog became known as *Grainger's Wholesale Net Price Catalog*, a name it would retain until 1989, when it became simply the *General Catalog*.

same time, the company continued the practices that Bill Grainger had initiated at Grainger's founding: fair treatment and true appreciation of the men and women who worked so hard to make the company a success.

A Slew of New Products

Though the number of items in Grainger's catalog in the early 1980s would seem small when compared to the number at the end of the decade, it was little wonder that *Forbes* magazine dubbed Grainger "a Sears Roebuck of industrial supply" in October 1982.[6] In the spring of 1981, *MotorBook* No. 359 offered over 450 new products, making it one of the largest issues in a decade, with more than 9,100 items. Another 600 items were added in 1981, 800 in 1982, and 900 more in 1983. Grainger's engineers carefully evaluated new products for performance, service life, and safety before offering them to customers. Products included greater-horsepower Dayton motors; screws, anchors, nails, staples, and other hardware needs for the remodeling and building industries; electronic cash registers, more-sophisticated phone answering systems, and calculators for smaller businesses; a larger selection of track, recessed, and emergency lighting; and energy-saving water heaters, furnaces, air conditioners, and motors.

Grainger's catalogs went beyond mere lists and descriptions. Each catalog educated customers by providing detailed product specifications, performance ratings, product comparison charts, and technical information on codes and industry standards—all in layman's terms.[7]

"The catalog's intent was to make it easy for the customer to buy," said Max Mielecki, who worked with the head of engineering, Emil Bahnmaier, to develop glossaries that explained the catalog's technical terms and tables. "It was the age of the do-it-yourself industrial, and the typical customer was proficient with his hands, but he might need a little technical assistance. We developed easier ways to order or to estimate what customers needed. Every time you put a piece of information about a product in the catalog, chances are you forestall about 2,000 telephone calls."[8]

As the number of pages in the catalog increased, Grainger printed fewer installments.

Not until the late 1980s did Grainger begin advertising in trade magazines. This print ad, which debuted in 1987, touts the reliability of Grainger's Dayton-brand industrial-duty motors and the ease with which the motors can be ordered.

In 1965, for instance, the annual number had been reduced from four issues to three, with the spring catalog (No. 314) containing 356 pages. By contrast, No. 359, in the spring of 1981, contained 988 pages. That was when Grainger began printing just two *MotorBooks* per year, in January and July, with supplements printed in April and October.

These supplements featured new products and seasonal items and grouped items by specific markets, including energy-saving, air-moving, and farm-duty commodities. Published in a color that accented the related *MotorBook* issue, the cover of each supplement featured an index divided into each category's major products. Because new product listings were presented more concisely within the supplements, clients could also review them more quickly.

Technological Developments

Grainger had never been shy about investing in its future, whether for new products, larger facilities, or more efficient processes. So it was little wonder that, technologically, the company was moving ahead in leaps and bounds.

In December 1980 the automated storage and retrieval system (AS/RS) was completed. Located on the south side of the Central Distribution Center, in Niles, the 84,000-square-foot AS/RS structure was equivalent to a 250,000-square-foot conventional warehouse and would now serve as the computerized headquarters of the distribution center's inventory. Ten stacking cranes controlled by computer could automatically store and retrieve the warehouse's 32,000 pallets of product. The automated system increased productivity and accuracy and provided better security. The distribution center in Niles performed so well, in fact, that in 1984 it received the American Material Management Society's Plant of the Year Award.[9]

"It took dogged determination to build that thing," said George Rimnac, vice president and chief technologist. "AS/RS technology was very, very new at that time, but Grainger had determined that it wasn't just going to build an AS/RS; it was going to build a really big one. We ended up having a very complex computer-controlled conveyor system, and getting everything to work was a big triumph for us."[10]

"For Grainger, the AS/RS was space-age technology," said John Slayton, who was hired as a project engineer for the AS/RS in 1979 and later became senior vice president for supply chain management. "The whole building operates with no one in it, and overnight we jumped into some highly sophisticated and automated equipment when the most sophisticated automation we had before that was an in-floor towveyor, which was like a carousel."[11]

By March 1983, Grainger had installed a computerized branch order processing and inventory control system (BOP/ICS) in all its branches. The internally developed BOP/ICS drastically enhanced customer service and branch productivity, improved inventory control, and made better use of warehouse space.[12]

Wiley Caldwell described how antiquated the old system of taking orders was.

Let's say Mary is talking to a customer on the phone and has 10 phone lines in front of her, and the customer says he'd like to buy three 2N743s. So Mary would put him on hold and push a squawk box speaker on her desk and say, "Hey, Joe out there in the warehouse, have you got any 2N743s?" Joe would have to walk clear across the warehouse to check availability, then come back and tell Mary. Now Mary has to remember which phone line she was on of those 10, because now she's picked up another customer. She hopefully goes back to the right customer and says,

By the end of 1980, Grainger had finished construction in Niles of its automated storage and retrieval system (AS/RS), which was a model of materials-handling efficiency.

"Yes, we have them," or "We have two, but we need to order the third," and writes the order down on a piece of paper to be entered into the computer manually by someone that night. Well, if another customer walked into the branch and wanted the same item, we had no way of telling that the item had been reserved for somebody else.[13]

Caldwell set up a task force to investigate how Grainger could improve its ordering system, which resulted in the BOP/ICS software being tested at the Toledo, Ohio, branch.

We went over a couple of days after it was running, and I remember the Toledo branch manager saying how much customers loved it. They were used to hanging on the phone waiting while we checked stock, and they couldn't believe how quickly we were able to check availability.[14]

By the end of the decade, all the branches would throw out order pads and embrace computer technology. Grainger, in fact, remained on the cutting edge of technological developments well into the next century.

Procession of Leaders

As the company grew, new management positions contributed to efficiency. In September 1981, Wiley Caldwell was elected to the office of executive vice president. Previously president of the Distribution Group, Caldwell would now oversee human resources, company planning activities, and both the Distribution and Manufacturing groups. Two years later, Caldwell was named Grainger's president. Jere Fluno, formerly vice president of finance, was appointed senior vice president and chief financial officer in 1981, while Richard Norman, previously vice president for marketing, was elected president of the Distribution Group. (Norman resigned in 1985, and Caldwell reassumed the mantle of president of the Distribution Group.) Also in September 1981, James Baisley joined the company as vice president and general counsel. And in 1983, John Rozwat became vice president for marketing to help Grainger improve its competitiveness in products and services.

Clockwise from left, Jere Fluno, vice chairman, Wiley Caldwell, president, and David Grainger, chairman of the board, made an impressive triumvirate of management, with each contributing unique talents to the success of the company.

As Grainger celebrated its triumphs, it also mourned the loss of its founder. On October 9, 1982, William Wallace Grainger died at his home in Wilmette, Illinois, just north of Chicago. He was 87 years old. Four weeks later, on November 7, Hally, his wife of 59 years, also passed away. Bill and Hally were survived by their son, David, their daughter, Barbara Tresemer, seven grandchildren, and five great-grandchildren.

Bill Grainger would be remembered not only as an engineer and businessman but also as a philanthropist. Before his death, an endowed chair was established in his name at the College of Engineering at the University of Illinois in Champaign-Urbana, where he had received his electrical engineering degree in 1919. And in 1949, Bill and Hally had established a charitable trust, later renamed The Grainger Foundation, Inc., which has donated tens of millions of dollars to hospitals, churches, schools, museums, and residences for exceptional children and needy adults.

Good Communications

With new management in place, in December 1982 Grainger launched a bimonthly newsletter called *Corporate Focus* for its corporate employees. The publication's premiere issue said it would reflect "the unique personality of our working environment" and support "the functions of the corporate organization."[15]

Grainger's Manufacturing and Distribution groups had their own publications. *The Doerr Way* represented Manufacturing Group employees, while the *Grainger Grapevine*, first published in December 1948, evolved into a new publication called *Network*, representing Distribution Group employees. The premiere issue of *Network*, in January 1983, announced that it had replaced the *Grapevine* "to more clearly reflect who we are and what we do."[16]

At that time the Distribution Group had 167 branches and 532 sales representatives. Employees aged 18 to 35 accounted for 70 percent of its workforce. *Network* catered specifically to these employees' needs, providing information about Distribution Group goals and programs, the work of various departments, economic news affecting the business, and general information about the group's network of sales representatives, branch employees, distribution centers, and Chicago support personnel.

In January 1983, the company put into writing its long-established operating principles. As David Grainger explained, "We are defining the context in which we want our managers to make future decisions. As we become larger and more complex, it is vital to communicate to all our employees what the company stands for so that we are all working toward the same objectives."

The company's principles were as follows:

Be committed to:
- *Superior service and satisfaction for all customers*
- *Mutually fair and responsible arrangements with vendors*
- *Fairness, dignity, and opportunity for all employees*
- *Competitive compensation and benefits*
- *Professionalism in all aspects of business operations.*

Operate for prudent growth while sustaining:
- *Historic or improved operating ratios*
- *Sound and conservative financial policies*
- *An attractive rate of return for the shareholders.*

Employ persons who:
- *Are qualified to discharge their assigned responsibilities*
- *Are dependable and loyal*
- *Have high standards and integrity*
- *Have empathy and concern for others.*

Be a good corporate citizen in the communities in which we operate.

Incorporated within company-wide policy manuals, these principles were also distributed to employees with their final 1982 paychecks and displayed on company bulletin boards and in newsletters, benefits binders, and handbooks.

For those who knew David Grainger, such operating principles were to be expected. Jim Baisley remembered that his first impression of the company when he joined as general counsel in 1981 was "a combination of class and integrity, much more so than the average company.... Everything was done first class, and the integrity had a lot to do with Dave Grainger and the way he conducted himself. If anybody did anything that was the slightest bit dishonest, you were out of there in a hurry."[17]

Greater Distribution

In its earlier days, Grainger had distributed products over the counter to customers or delivered products directly. But as more customers began ordering an increased number of products, Grainger had to find fast and efficient ways to order and store its massive inventory and ship it to its nationwide network of branches.

Even after adding the AS/RS facility to the Central Distribution Center in Niles, Grainger still needed more warehousing space. In January 1983, the company opened another regional distribution center, this one with a flow-through design. Sprawled over a 180-acre site in Kansas City, the 1.4-million-square-foot facility more than doubled Grainger's available space for stock. With 115 loading docks for trucks and six miles of conveyors to automatically move merchandise, the center served 92 branches in 27 states and greatly enhanced Grainger's ability to serve customers throughout the United States.[18]

"We were growing very quickly," said Caldwell. "But we knew that if we were going to put one of these 1-million-plus-square-foot warehouses up, we'd better be there for the long term. It was a leap of faith to build the one in Kansas City, but it turned out to be very successful."[19]

Partnership for Success

Following the precedent of years past, Grainger continued offering incentives and benefits that kept its workforce happy and productive. During the 1940s and 1950s, company benefits had been considered extras. In the 1960s, awareness of the importance of benefits was growing, but it wasn't until the 1970s that job candidates began seriously evaluating benefit plans along with compensation when considering employment offers. To many, benefit levels were now as important as compensation, and in this area Grainger had a proven track record of always being ahead of the game.

Based on a 1979 in-house review of the company's benefits program, outside standards, and employee feedback, Grainger announced significant changes to its benefits package in December 1980. These included the introduction of dental coverage and a short-term-disability plan, updated surgery payment methods, and an increase in the number of paid holidays to 11.

In May 1983, Grainger introduced a new employee stock ownership plan (ESOP) to reinforce the "partnership for success" policy that it had initiated with the profit sharing trust 42 years earlier. All employees with at least three years of service who worked a minimum of 1,000 hours per calendar year would now share equally in the ownership of Grainger common stock, purchased with funds contributed by the company to the ESOP trust. Grainger's contribution in 1982 of approximately $800,000 bought from three to four shares of stock per person. (At this time, the company had around 14 million shares of outstanding common stock, all traded on the New York and Midwest stock exchanges.) Employees didn't hold the actual stock certificates for the shares in their accounts, but they were entitled to all shareholder privileges, including receiving annual reports and financial updates and voting in the annual election of directors.

The employee stock ownership plan was established to take advantage of federal legislation authorizing a tax credit for the funding of ESOPs. The Tax Reform Act of January 1987 repealed that credit, however, so Grainger ceased contributing to the ESOP. To replace it, in March 1987 Grainger

---

Opposite: Grainger's exclusive President's Club was established in 1986 to recognize branch managers and sales personnel who displayed outstanding achievement. The 1988 inductees are shown, as well as President Caldwell (front row, fourth from left).

implemented an employee stock purchase (ESP) program, which enabled employees to purchase company stock without paying brokers' fees. Grainger assumed the cost of brokerage commissions paid to Merrill Lynch on all payroll deduction purchases and on additional company stock added to employees' accounts through the dividend reinvestment plan. The ESOP would be terminated on December 31, 1989, with participant account balances being transferred to a newly created investment option of the profit sharing trust.

To recognize outstanding achievement by branch managers and sales representatives, Grainger established the President's Club in 1986. The first 12 inductees traveled to Chicago in March 1987 to be recognized for their 1986 successes. They were awarded prestigious red-stone rings—the highest honor bestowed on members—along with business cards and leather briefcases that had been personalized and engraved with the President's Club logo.

Superior Sales

While Grainger beefed up employee benefits, it also took steps to increase sales. In 1984, the Distribution Group participated in 12 trade shows to increase its market penetration and visibility. That year Grainger created an export sales department to increase its penetration of foreign markets. With eight sales representatives focusing exclusively on foreign sales that year, export sales reached a new high of $10 million.[20]

In addition, Grainger began a new sales strategy in 1984 to compete with original equipment manufacturers (OEM) by selling a total package via its nationwide service-and-supply network. (Grainger's OEM accounts were those that included one or more *MotorBook* items as components in a finished product.) Grainger's specialized sales representatives worked as liaisons between the OEM purchasing, engineering, and design personnel on one side, and Grainger's vendor, merchandising, and engineering departments on the other.

June of 1984, meanwhile, was Grainger's first $100 million sales month, thanks to an increased emphasis on customer service, new products, larger branch inventories, a more aggressive branch/sales representative effort, the new distribution center in Kansas City, and the automated storage and retrieval system at Niles. The milestone sales record meant that Grainger had outpaced the gross national product's growth rate by a ratio of

four to one, surpassing its typical ratio of three to one. Grainger celebrated on June 27 by providing employees at the branches, distribution centers, and group headquarters with coffee and cake, also distributing "We Made It Happen" posters that listed the names of all Distribution Group employees.

Grainger reached an even bigger milestone on December 7, 1984, when for the first time it surpassed $1 billion in annual sales, which had increased 20.4 percent from the previous year to $1.06 billion. Each of more than 6,000 full-time employees received a bonus of $50, while 17-foot banners announced "Thanks a billion for making '84 our first billion dollar year."[21]

Improving Operations

As Grainger grew, so did its need for a highly specialized workforce and more modern distribution operations, especially as the company continued to add more products. Grainger turned its attention inward to improve efficiency, particularly for its distribution arm.

The company was able to improve service and sales while lowering inventory costs of its branches with an internally developed computer system called LINQ (local inventory inquiry).

Installed in all of Grainger's branches by 1985, LINQ shared product information among branch offices, allowing branch employees to quickly locate and order items that might not be in stock at their own branch.[22]

At the same time, Grainger expanded its telemarketing program to shave costs associated with personal sales visits to its smaller accounts, allowing sales representatives to concentrate on major customers. Using computers, Grainger was able to build customer profiles based on how many and which items its 840,000 customers typically bought. (At that time, Grainger had estimated that 80 percent of its customers bought less than $300 worth of products a year.) The marketing department could then target customers' needs by mailing specific marketing material and following up with phone calls. By the end of 1985, Wiley Caldwell said that Grainger had doubled its rate of sales to smaller-volume customers. "We found telemarketing to be highly controllable," he said. "It permits us to penetrate markets and target whole industries in a matter of days with a minimum of staff."[23]

Grainger also implemented several formal training programs to ensure that its sales representatives and branch personnel continued their excellent customer relations while amassing greater sales volumes. In early 1985, the branch operations and training and development departments launched a new customer contact program focusing on

Grainger's office headquarters building in Skokie, Illinois, was completed in 1978.

improved communication of product information, telephone skills, and suggestive selling. After Grainger tabulated 158 branch profiles, it drew up a "top 10" list of skills and qualities required for branch personnel. The company wanted employees with empathy; a good knowledge of product; courteous behavior; a friendly personality; good communication and listening skills; speed; professional appearance; knowledge of policy, procedures, and resources; a positive attitude; and patience.

Also in early 1985, the *MotorBook's* name changed to *Grainger's Wholesale Net Price Catalog* to better reflect its diverse line of products. "Motors and that category of products were only about 35 to 40 percent of our business," explained Caldwell, "and the percentage was getting lower as we added new products. We had an advertising agency do a bunch of different Grainger logos for us, and we decided which one we liked, and that became the logo on the catalog."[24]

To enhance employee training and communications, Grainger opened its training and communication center (TACC) at the south end of the company's headquarters in Skokie in December 1986. The TACC comprised three meeting rooms equipped with audiovisual gear and a state-of-the-art video production suite. With its convenience and reliability, the facility enhanced employee and management training.

Focus on Distribution

During the first half of the 1980s, Grainger invested in its Manufacturing Group by expanding capacity, lowering costs, and providing better products and services to customers. In 1981, the Manufacturing Group added a computer-aided design and manufacturing (CADAM) system that improved engineering methods and productivity and implemented state-of-the-art computerized technology.[25] Two years later the group moved its headquarters to a new, 60,000-square-foot building in Cedarburg, Wisconsin, while its previous headquarters at the Cedarburg manufacturing facility was renovated for manufacturing space and expanded administrative offices. Also that year the group formed a new unit, Dayton Industries Division, which manufactured industrial products such as fans, blowers, air compressors, and pumps for the Distribution Group at a plant in Northbrook, Illinois. That year a new facility was added in Lenexa, Kansas, to expand production of air-moving products.[26]

In 1985 the company decided to hike its capital expenditures and concentrate on its core competency of distribution. The decision resulted in part from the decline in sales and earnings the company had experienced in 1982. Grainger began to think more strategically about long-term success, and one

of the conclusions company officers reached was to sell off Manufacturing.

"Some analysis suggested that we were bringing a distributor's logic and risk preferences and ways of thinking to a business that didn't respond well," said Andy Thomas, director of strategy. "For example, we were seriously underinvesting in long-term R&D, and at the time, electric motor technology was evolving pretty quickly."[27]

"We were smaller than GE or Emerson, and we couldn't afford the R&D that those companies could afford," explained Caldwell. "Then when GE began making the Dayton label for us, we knew we didn't need Doerr Electric anymore. The decision to sell Doerr required an understanding of our overall business and strategy, which is to say, 'Why should we be in manufacturing when we didn't need to be?'"[28]

To that end, on January 10, 1986, Grainger completed the sale of Doerr Electric Corporation, its electric motor manufacturer, to the St. Louis–based Emerson Electric Company.[29]

"We thought we should concentrate on what we knew best, which is distribution," David Grainger told *Crain's Chicago Business*.[30]

As part of the Doerr Electric deal, Grainger stipulated that Emerson couldn't change the former Grainger employees' benefits program for three years, and those employees in Grainger's profit sharing trust automatically became fully vested.[31]

The divestiture of Doerr gave Grainger a onetime gain of $11.3 million in the first quarter of 1986.[32]

After the sale of Doerr Electric, Dayton Industries began operating as a separate company division. But its fan and blower assets were bought by Emerson in February 1989. Just a few months later, Grainger sold the remaining assets of Dayton Industries, including the Northbrook plant's fabrication and assembly operation of pumps and compressors, to the Scott Fetzer Company of Westlake, Ohio.

In 1986 Grainger formed another new division, the Specialty Products Division, which allowed the company to expand parts sales, increase its market share for lower-demand wholesale catalog goods, and enter specialized markets where it could distribute directly.[33] From a central location in suburban Buffalo Grove, Illinois, Grainger sold tens of thousands of replacement parts and supplied products such as power tools and electric motors that couldn't be sold profitably through its branches. In 1986, the division issued two separate sales catalogs—one featuring test and measurement instruments and the other, safety equipment.[34]

Overall, Grainger's focus on distribution resulted in a significant boost in 1986 sales, which

In 1986, Grainger formed the Specialty Products Division and released two catalogs, one for test and measurement instruments and this one for safety equipment.

Donald Bielinski began his Grainger career in 1972 as accounting supervisor and rose to group president in 1997. He retired from Grainger in 2001.

rose 6 percent from the previous year to $1.16 billion. Earnings rose 20 percent to $86.1 million.[35] The following year, Grainger again achieved record sales and earnings. By that time, Grainger's catalog offered more than 19,000 products, while its Specialty Products Division offered 3,600 items.[36]

New Talent

Some key appointments were made within Grainger during the mid-1980s. In 1985, David Barth, previously vice president and treasurer, was elected vice president for planning and development, a newly created position. Donald Bielinski moved from his position as controller of the Distribution Group to vice president and controller. Robert Pappano shifted his role from vice president and controller to vice president and treasurer. By 1989, Bielinski would be elected vice president and chief financial officer, and Paul Wallace would be named vice president and controller.

Meanwhile, Wiley Caldwell set about putting a succession plan in place. "I knew I was going to retire soon, so Dave [Grainger] and I brought the issue up with the board. As a result of that concern, we hired Dick Keyser."[37]

Richard L. "Dick" Keyser became vice president of operations for the Distribution Group in the fall of 1986. Keyser had most recently been president of Hycalog division of NL Industries in Houston. Before that, he had earned a B.S. in nuclear science from the U.S. Naval Academy and an M.B.A. from Harvard Business School and spent 13 years with Cummins Engine Company. Keyser was 43 when he began at Grainger and would become the company's president and chief operating officer in 1994, CEO in 1995, and chairman of the board in 1997.

A year after Keyser was hired, P. Ogden Loux, who had been a financial manager in various General Electric businesses, joined Grainger as vice president and controller of the Distribution Group.

The company mourned the loss of another of its early leaders when on April 19, 1987, Elmer R. Slavik, son of Elmer O. Slavik and father of James Slavik, died at the age of 60 from a brain tumor. E. R. Slavik had served as a member of Grainger's board for 21 years. In the year prior to his death, he had decided not to run for reelection to the board and began grooming his son James for the position he would vacate. The younger Slavik was just finishing his M.B.A. when his father fell ill and realized it was time to start thinking about a successor. James Slavik was elected to the Grainger board of directors at the annual shareholders' meeting just 10 days after E. R.'s death.

"I felt a lot of responsibility," Jim Slavik said years later. "We didn't have a long period of time for him to teach me—not like when you know you're going to retire in three or four years. So I had to jump in with both feet, and I wasn't sure I was ready for that. But the board was fantastic. All of the people on the board and the people in management helped bring me up to speed, and I spent my first year just learning. It was a steep learning curve, but it has always been very rewarding."[38]

Convenience Is Key

While Grainger continued investing more capital in its distribution system to serve its customers with greater efficiency, the company initiated an extensive marketing study in 1986 that turned up some very interesting numbers. Though it had doubled its sales over the last decade and was bigger by far than any of its competitors, W. W. Grainger, Inc., held only 2 percent of the estimated $70 billion to $90 billion wholesale market. "That's kind of embarrassing to us, that we don't have more," David Grainger told the *Chicago Tribune* in July 1987.[39]

John Schweig was a member of the strategy consulting firm Bain and Company, which had been hired to help put Grainger on track. "Grainger was a really interesting story," Schweig said. "If you just stare at a chart that looks at its revenue from the day Bill Grainger sold his first motor to today, you see this very steep and steady graph of growth. But from 1979 to 1985, all that growth was basically inflation, and when you draw the same chart without inflation, you see that Grainger was no bigger in 1985 than it was in 1979."[40] Schweig later became Grainger's senior vice president for business development and international operations.

The company, it seemed, had been neglecting enormous potential in major metropolitan areas, where management had assumed it had enough branches to thoroughly penetrate the market.

With such a huge market and fragmented competition, it was time for Grainger to take advantage of the opportunities.

The market analysis showed that most of Grainger's customers had fewer than 100 employees. With such a small infrastructure, these customers considered a convenient branch location to be more important than any other factor, including product availability, variety, and price. Most of Grainger's branch sales catered to walk-in business, and customers were reluctant to travel farther just to do business with Grainger if there was a competitor closer by.

"We learned that there is nothing unique about the customers we sell to," said Andy Thomas. "There's nothing unique about the products we sell either. But we learned that the circumstances in which customers come to us are unique. It's about speed, and it's about convenience, and we do it in a highly proprietary and unique way that we've worked on for 50 years. That's hard for anyone else to duplicate and hard for people to compete with."[41]

Grainger's executives took the marketing study to heart and in January 1987 announced they would expand the network of 200 branches by at least 50 percent in the next three to four years, saturating metropolitan markets with easy-access branches. Over the previous decade, Grainger had expanded at an average rate of six new openings a year.[42] Under the new program, sites for new branches would be selected using a demographics model that identified the location, number, and type of customers by zip code. Robert Thrush, currently vice president of sales, was responsible for developing the selection process model. He noted in a 1997 interview that, though "it's been updated and tweaked over the years, that model lives on today."[43]

New York, Detroit, Chicago, and Los Angeles were slated as the first metropolitan areas for expansion. In 1987 alone, Grainger opened 16 new branches in the New York City area, five in Detroit, and 15 more in smaller cities. By the end of the year it opened the first of 11 planned branches in the Chicago area.[44] Wiley Caldwell told shareholders at the annual meeting in April that he anticipated "at least a hundred new branches over the next three or four years of rapid but carefully controlled growth."[45]

To power that growth, Grainger committed up to $40 million for capital improvements in 1987, compared to $30 million the year before. "This is a new era of growth for the company," Caldwell said. "We are reinvesting in our core business."[46] And because Grainger traditionally kept a conservative balance sheet with long-term debt at only 3 percent of capitalization in 1987, it was able to support the expansion program with internally generated funds.[47]

Analysts liked what they saw, predicting that Grainger's saturation of key markets would certainly pay off. "If quarterly earnings suffer, so be it," said Caldwell. "Too many companies have managed to the short term."[48]

In June 1987, Grainger's stock reached a new high on the NYSE, trading at $56 a share.[49] By February of the next year, the stock had soared to the $70 range.[50] "It's an extremely well-managed enterprise," said Walter Kasten, senior vice president at Blunt Ellis & Loewi. "Plus they're not afraid to employ long-term strategies at the sacrifices of a quarter or two."[51]

"We believe W. W. Grainger offers an exceptional investment opportunity," wrote an analyst with Robert W. Baird & Company. "Long recognized as a high-quality company on the basis of its A+ financials and dominant market position, [Grainger] has become a growth company.... The business [is] there to be had and Grainger [has] an excellent chance to garner it, if it [can] get its broad product line closer to the customer.... The list of underserved major cities is long and promising."[52]

A Preeminent Distributor

In 1988, Grainger outlined a vision statement designed to accelerate the company's growth through intensified marketing efforts. Among the statement's directives were the desire "to be a preeminent broadline [complete-product-line] distributor of equipment, components, and supplies" and to "attain leadership

Opposite: Grainger's 1983 board of directors at the regional distribution center in Kansas City. From left: Elmer R. Slavik, Harold Smith Jr., Jere Fluno, George Baker, David Grainger, Graydon Ellis, Kingman Douglass, Edward Duffy, Wiley Caldwell, and Robert Elberson.

positions as a specialty distributor." At the same time, the company sought recognition "as a source for excellent value and service," "an outstanding place to work," and "an efficient, reliable marketer" of its vendors' products.

In truth, Grainger had already achieved most of this vision by the time it was spelled out, but David Grainger and the company's other leaders wanted to ensure that they did not stray from their original intent. The actions the company took thereafter were guided by that vision.

While Grainger expanded its branch system, it also fortified its program to train branch managers at a faster rate, ensuring that it did not spread its management too thin. New branch managers underwent an intense 10-month on-the-job training program that brought them up to Grainger's high standards.

Meantime, further marketing studies showed that, like most MRO (maintenance, repair, and operating supplies) customers, the bulk of Grainger's customers prized speed and convenience over price. The company thus redefined the market size of its core business from $70 billion–$90 billion to $50 billion.[53] But Caldwell noted that "Our customers want more than convenience. They want zero mistakes on specifications, price, quantity, availability, item numbers, and so on.... We must provide the fastest possible service at every point of contact with the customer."[54]

With half of Grainger's sales made over the phone by customers ordering from the catalog, the company beefed up the training of those who handled the phones.[55]

In 1987, warehouse employees began using handheld terminals that communicated with the local computer system. The new technology not only improved branch inventory control but also improved accuracy and saved time on labor. Grainger also upgraded its computer system with new IBMs in all its branches. The expandable

Wiley Caldwell, president, inspects the bin storage section of the main Chicago warehouse. Caldwell played a major role throughout much of Grainger's "continuing era of growth."

computer system halved the average time it took a telephone salesperson to process an order.[56]

Moreover, the company refurbished its branches to make them more aesthetically pleasing and began a salesroom program to trigger impulse buying by displaying popular goods at the branches' sales counters.[57]

An article in *Forbes* magazine praised Grainger's efforts:

These days, service is so swift that the average customer at Niles is in the branch for only seven minutes, about the amount of time it takes to find a sales clerk at Builders Square. Grainger customers can get same-day shipping on any item, and three-hour emergency delivery service is available for an extra fee.[58]

Grainger's unprecedented facilities growth continued into the last years of the decade with 55 new branches in 1988 and 25 in 1989. The new branches were located mainly in or near metropolitan areas that had a large commercial and industrial base and a correspondingly larger number of customers.[59]

On September 18, 1989, the company celebrated the opening of its 300th branch, in Daytona Beach, Florida, and that November the company's first offshore branch began operations in Honolulu, Hawaii. That was followed by a branch opening in Alaska in mid-December. By the end of the year Grainger had a total of 311 branches, located in all 50 states of the union.

To help replenish the inventory to those new branches, Grainger opened a third distribution center, in Greenville, South Carolina, in February 1989. Operating under the mantra "Do it right the first time," the 1.1-million-square-foot facility provided service to 95 branches in the southeastern and northeastern United States and boasted 126 dock doors and parking for more than 300 trailers.

Grainger also began a national advertising campaign in 1987 with the theme "There is more to Grainger than you think" and established a toll-free after-hours emergency telephone number.[60] A marketing study in 1989 showed that 95 percent of purchasing and maintenance people had heard of Grainger, but only a paltry 5 percent really knew what the company offered. Thus Grainger intensified its training efforts so that its 900-plus sales force could guide customers through its enormous product lineup. The company also revamped its catalog, which by 1989 listed more than 22,000 different items, to make it more user-friendly and manageable.[61]

Sitting Pretty

As the company maintained its aggressive strategy to be the leader in the market, Wiley Caldwell began referring to what he had termed Grainger's "new era of growth" as a "continuing era of growth." Still financially flush, Grainger had been able to avoid debt despite its rapid expansion, and sales and earnings were swelling with each fiscal quarter.[62] Grainger's 1989 sales increased 12.5 percent from the previous year to $1.7 billion, while earnings rose 9.9 percent to $119.5 million.[63]

It was no wonder that in 1989, *Fortune* magazine listed Grainger as No. 56 in its *Fortune* Service 500 listings.[64] The company had worked hard during the decade to gain such glowing status. With clear goals and excellent leadership, W. W. Grainger, Inc., was more than ready to take on whatever challenges the 1990s had to offer.

David Grainger shakes hands with Wiley Caldwell during Caldwell's Service Recognition dinner, which took place a year after his retirement in 1992.

CHAPTER TEN

STRATEGY FOR SUCCESS

1990–1995

IN 1990, *FORTUNE* MAGAZINE listed Grainger, with earnings of $126.8 million and sales of $1.9 billion, on its *Fortune* 500 list, marking the beginning of a new era in Grainger's development. The company had become a premier distributor of maintenance, repair, and operating (MRO) products, providing one-stop shopping and lower total procurement costs for a growing number of businesses that depended on Grainger to keep their operations running smoothly. While other wholesale distributors were reducing inventory due to the weakened economy, Grainger was growing as never before, thanks to its rapid expansion program, technological upgrades, and focus on customer service. Yet Grainger's journey was far from over. Now the company was faced with a new challenge: how to gain market share in an increasingly competitive, fast-paced environment while maintaining its high standards and improving customer satisfaction.

Management Changes

In the midst of Grainger's development, the company's new management continued to bring leadership talent to the forefront. Effective January 1991, Richard Keyser was named president of Grainger Division, formed in 1990 with the Specialty Distribution Group. Keyser had recently served as executive vice president and general manager of the division. The following year, Lee Flory, who had been vice president and secretary, retired after 42 years of service, while Jim Baisley, vice president and general counsel, took on the additional role of corporate secretary.

On July 31, 1992, Wiley Caldwell retired, passing the presidential reins to David Grainger. Richard Keyser, meantime, was elected to the new position of executive vice president with responsibility for both Grainger Division and the Specialty Distribution Group. In response to the increasing size, complexity, and diversity of the company and to ease the issue of succession (David Grainger had turned 65 in 1992), the company created the Office of the Chairman, shared by Jere Fluno, then vice chairman; David Grainger; and Richard Keyser. Any two of the three men were empowered to make a decision in the other's absence.

In 1992, Grainger acquired Lab Safety Supply, which sold 13,000 personal and environmental safety products nationwide. By 1995, Lab Safety's award-winning *General Catalog* had reached 1,100 pages and offered nearly 34,000 products.

"The Office of the Chairman was unique because we had three people who worked well together," said Fluno. "We all had our certain strengths and certain weaknesses, and we complemented each other."[1]

Richard Keyser was elected president and chief operating officer in 1994. The following year, at age 52, he was appointed chief executive officer. David Grainger remained chairman of the board. As the company transitioned to Keyser's leadership, it dissolved the Office of the Chairman.

National Accounts

Traditionally, Grainger's business had focused on distributing products to resellers, who then sold the products to the end user, in many cases a large company. But that was about to change as Keyser shifted the company's strategy to dealing more directly with end users. "We were using other resellers in the marketplace to get to more people," explained Micheal Murray, retired vice president of administrative services. "We were literally a distributor selling to a distributor that was selling to the end user."[2]

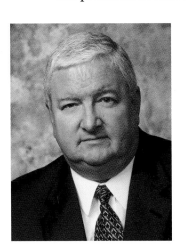

The reseller approach had worked since 1927, but by the 1990s it was choking the company's growth. In effect, the resellers were keeping Grainger from dealing directly with large customers. "We were at the point where the growth was starting to top out, and all of our branches had been geared to sell to other people who sold to customers," said Wesley Clark, who became group president in 1997. "We had lost touch."[3]

Mike Murray began working at Grainger the day after he graduated from high school. He worked part-time while he attended college and by the age of 21 had become an assistant branch manager. He retired as Grainger's vice president of administrative services in 2001, having served 37 years.

The customers Grainger had lost touch with were the large corporations, who often saw Grainger not as a branch-based business but as a world-class distributor of product. Grainger's business distribution model had been designed and developed to support the needs of small and medium-sized customers, but large corporations wanted different capabilities.

Grainger could now boast a facility within 20 minutes of 80 percent of the U.S. population. Keyser and the company's other leaders realized that Grainger had a superb platform from which to distribute product on a national rather than local level. The company had in fact already made the strategic and innovative decision to form a national accounts department that focused on larger customers with multiple locations. By 1989, Grainger served approximately 150 national accounts, including American Airlines, McDonald's, Coca-Cola, Delta Air Lines, and United Parcel Service, and that number was steadily growing.

The national accounts program, initially run by Mike Kight, grew significantly in the early 1990s. By 1995 it served 400 customers and generated $668 million in sales.[4] Local sales representatives and national account specialists worked to simplify the purchasing needs of these customers' various locations through inventory reduction; electronic commerce; and consistent procedures, products, and pricing—reducing customers' overall cost of acquiring equipment.[5]

New Businesses

With the national accounts program growing, Grainger reorganized operations to specialize in fulfilling the needs of particular customer segments and to provide more value to employees, suppliers, and customers. Two main units resulted: Grainger Division and the Specialty Distribution Group.

What had been known as the Distribution Group became Grainger Division, representing the core business of more than 32,000 catalog items. Grainger Division stressed speed and convenience. It focused on offering a wide array of maintenance and repair products to serve the immediate needs of contractors, property management companies, hotels, manufacturers, and institutions such as schools, health care facilities, and government agencies.

CHAPTER TEN: STRATEGY FOR SUCCESS

As part of its advertising campaign, Grainger used billboards to announce new branch openings. This billboard announces the opening of a branch in Springfield, Arkansas, in 1991.

The businesses that made up the Specialty Distribution Group, on the other hand, focused on planned purchases. Grainger's large national account customers had begun looking for distributors that could provide a low-cost model for planned purchases. Many of them knew what they needed and when they were going to need it, so they were no longer willing to pay Grainger for the cost of speed and the convenience of having a nearby branch. These large customers were also asking Grainger for products they couldn't find in its catalog. With its large customers' needs in mind, Grainger decided to invest more heavily in specialty products.

The Specialty Distribution Group consisted of Bossert Industrial Supply, formed after Grainger made a series of acquisitions; Jani-Serv, a greenfield start-up company that supplied janitorial supplies; Parts Company of America, which handled 80,000 brand-name spare and replacement parts for industrial and commercial equipment; and Allied Safety, which distributed personal and environmental protection products and equipment. Allied was formed after Grainger purchased two safety supply distributors in 1990.

The companies existed as separate business units for five years, but in 1995 Grainger decided to begin integrating Bossert Industrial Supply, Jani-Serv, and Allied Safety into the core business. The primary reason for getting into the specialty businesses no longer existed, for Grainger's national accounts program had been thriving. "Grainger was getting better and better at selling on its own to these large customers," noted John Schweig, "so the specialty distribution businesses were no longer necessary."[6]

"At the end of the day, we were not really effective in getting into those other niche markets," added group president Don Bielinski, who was charged with the rather onerous task of blending the support functions of each of the three specialty businesses into the core business.[7] For the first half of the decade, however, the units operated as separate businesses and enjoyed their own share of successes.

Bossert Industrial Supply

Wanting to take advantage of businesses' planned purchases, Grainger had been considering its options for acquisitions for some time. In October 1989 the company had bought Vonnegut Industrial Products, its first acquisition in 17 years. Vonnegut's industrial supplies focused on production consumables for manufacturing operations and gave Grainger an added niche as a distributor of planned purchases. Founded in 1852 and headquartered in Indianapolis, Vonnegut had eight branches in Indiana and Ohio. It was a leading regional distributor of cutting tools, abrasives, safety supplies, and other products required in factories and production facilities.[8]

Grainger sought to broaden its new base by acquiring other industrial supply companies. In 1990, it purchased Bossert Company, of Kansas City, Missouri; C. L. Gransden & Company, of Dearborn, Michigan; Mansco-Lakeshore, of Grand Rapids, Michigan; W. M. Pattison Supply Company, of Cleveland, Ohio; SMS Supply Company, in Chicago; and the Satterlee Company, of Minneapolis, Minnesota.

These companies, combined with Vonnegut, formed Bossert Industrial Supply. With 20 locations in 1992, Bossert served customers in 12 states—primarily in the Midwest—and offered production consumables such as abrasives and cutting tools. Bossert sales representatives called on customers seeking manufacturing rather than maintenance products. During 1992, Bossert began forming partnerships with high-end suppliers to better serve its customers,[9] and the following year it introduced its first catalog, which focused on cutting tools and abrasives and featured 5,800 items.[10]

As Bossert widened its target market, its products would be sold and stocked by Grainger's core business.

Bossert eventually integrated with Grainger in September 1996.

Sanitary Supplies

Unable to find any appropriate sanitary supply companies to acquire, Grainger formed its own. Jani-Serv was launched in October 1990. The new firm distributed a complete range of sanitary cleaning supplies and equipment such as chemicals, paper goods, and accessories.

Headquartered in the 5959 Howard Street building in Niles, Jani-Serv initially established distribution centers in Chicago, Atlanta, and Philadelphia, each between 35,000 and 40,000 square feet. Distribution centers in Dallas and Los Angeles were added later. Jani-Serv's initial inventory totaled 1,200 items and targeted nationwide users of sanitary supplies and equipment who planned their purchases in advance and shopped around for price and quality. These customers included contract cleaning services, multisite manufacturers, hotels, and hospitals.[11]

In 1991, an acquisition opportunity arose at Ball Industries, a Los Angeles distributor of sanitary and janitorial supplies. Established in 1932, Ball served building maintenance contractors, producing more than 200 Ball-branded cleaning products and distributing sanitary products and equipment. With its own truck fleet, Ball was able to provide emergency and same-night deliveries.

Peter Torrenti, a manager at Ball when Grainger acquired it, noted that Ball staffers were impressed with Grainger's capability. "We knew Grainger was a financially sound company and was very, very well managed," Torrenti said. "They treated our people very fairly, improved our benefits programs, our training, education, and development, and our career pathing opportunities."[12] Torrenti later became a company vice president.

Already established in southern California when Grainger acquired it, Ball expanded into northern California in 1992 with a new facility in the San Francisco Bay area. The following year, Jani-Serv began selling Ball-name products.[13]

Meanwhile, Jani-Serv continued adding value. As a commitment to its customers, it established a Service First program, which guaranteed fast, accurate, centralized order processing and offered a 20 percent discount on catalog orders it could not ship the same day. In 1992, Jani-Serv supported more than 4,600 customers in 46 states.[14]

In 1993, Grainger combined its sanitary supply businesses into Grainger Sanitary Supplies & Equipment (GSS&E). The new entity took advantage of Grainger Division's nationwide sales and distribution network. In essence, GSS&E allowed Grainger to offer products of Jani-Serv and Ball, two regional companies, through its nationwide branch outlets. GSS&E became the nation's second-largest sanitary supply distribution business and by the end of 1994 added more than 1,500 sanitary products to Grainger's *General Catalog.*

Grainger had established a significant foothold in the janitorial segment with a national distribution network to back it up. But, Bielinski noted, the separate business unit didn't succeed because Grainger had yet to fully appreciate how to deal with large customers, which tended to be the target buyers for its sanitary supply business.[15]

According to Wesley Clark, who was hired in 1992 to solve the problems in the janitorial supply company and later became president and chief operating officer, "Customers were only interested in our [janitorial] services because we were associated with Grainger. We had to use the Grainger name to get into their accounts, and as soon as we did that, then customers were confused as to why they couldn't buy the same Grainger products from us as they could from Grainger. Basically, the company had its own warehousing and redundant overhead structure, so I came forward and asked that we integrate it into the base business."[16]

Safety Supplies

Grainger formed Allied Safety in 1990 after acquiring Allied Safety Supply Company, of Tucker, Georgia, and Jones Safety Supply, of Roanoke, Virginia. Operating from eight branches in five southern states in 1990 and featuring its own catalog, Allied Safety distributed industrial and environmental safety products such as respiratory systems, protective clothing, environmental cleanup products, and other safety equipment.

Allied's business got a boost when, in January 1992, Grainger acquired the Seattle-based Rice Safety Equipment, which had 1990 sales of $24 million. Rice Safety was a leading supplier of safety equipment in the Pacific Northwest with locations in Seattle and Spokane, Washington; Portland, Oregon; and Anchorage, Alaska.

Though Grainger had distributed safety products before the acquisitions, Fred Loepp, who joined Grainger in 1991 to head the safety distribution business, said the acquired companies had more technical knowledge of safety supplies. "Grainger was considered a secondary supplier of safety, not a primary supplier," Loepp said.[17] But that technical knowledge was important to Grainger, which was committed to serving

To improve service to customers and accelerate national expansion of its sanitary supply business, Grainger in 1993 combined Jani-Serv and Ball Industries to form Grainger Sanitary Supplies & Equipment (GSS&E). Later, GSS&E would be integrated into Grainger's core business.

GOOD CORPORATE CITIZEN

WHILE GRAINGER CONTINUALLY IMproved its operations, it also strove to be a good corporate citizen by staying environmentally responsible and by giving back to the communities it served.

For example, Grainger in 1991 became the first major distributor to join the Green Lights Program of the Environmental Protection Agency (EPA). The program sought to save energy, conserve resources, and reduce pollution through use of energy-saving lighting technologies. Grainger began installing such lighting in all its facilities shortly after joining the program. For hazardous mercury-bearing fluorescent and high-intensity discharge lamps and batteries, the company provided a national recycling program that complied with the EPA's Universal Waste Rule.

Grainger's Matching Charitable Gifts program matches employees' charitable contributions three-for-one up to a total of $7,500 each year. The company also donates products to many nonprofit organizations, including the National Association for the Exchange of Industrial Resources (which distributes donated overstock inventory to nonprofits and schools) and Educational Assistance Ltd. (which gives excess inventory to colleges and universities in exchange for scholarships for needy students).

The company also supports a number of charitable organizations such as i.c. stars, which provides technology training to high school graduates in inner-city Chicago with the goal of helping them find secure leadership roles in corporate America. In addition, Grainger contributes to the National Minority Supplier Development Council's Advanced Management Education Program, which enhances skills of minority business owners to help them grow their businesses.

In addition, Grainger helps the communities it serves by acting swiftly to stock its branches with emergency supplies during storm and flood seasons. When hurricanes Hugo and Andrew hit, for example, Grainger employees rushed to the emergency areas, and branches stayed open as long as necessary to ensure that items such as power generators, pumps, flashlights, and tarps were in ready supply.

Grainger was on hand again to help victims of the terrorist bombing of the Alfred P. Murrah Federal Building in Oklahoma City on April 19, 1995. Within hours of the bombing, Grainger coordinated with local authorities and sent branch and sales employees to set up a makeshift Grainger branch inside a parking garage located a block from the disaster site. Situated at a command center near the Red Cross station, the temporary branch responded to requests of the Federal Emergency Management Agency and the FBI. It remained open 24 hours a day for a week to supply emergency products such as boots, shovels, hard hats, batteries, flashlights, wire cutters, and respirators. On May 16, during a ceremony in Oklahoma City, Grainger presented a $10,000 check to a relief fund for victims of the tragedy.

After the terrorist attacks of September 11, 2001, Grainger donated $1 million in cash and emergency supplies to help rescue-and-recovery teams in New York. Thanks to Grainger's speedy distribution network—the company was able to mobilize within minutes after the attack—hard hats, steel-toed boots, respirator masks, safety glasses, gloves, and other supplies were quickly delivered where they were so desperately needed. Police escorted Grainger's delivery as it supplemented local inventory with products from its New Jersey distribution center and 20 other branches. In addition, a Grainger branch about two miles from "Ground Zero" remained open 24 hours a day for several weeks to quickly provide emergency supplies to response teams. (See sidebar on page 132 for more information.)

customers as the "Source for Safe Solutions." In 1992, for example, Allied Safety professionals conducted 25 seminars for health and safety managers and industrial hygienists to help them comply with Occupational Safety and Health Administration regulations. The seminars also presented other topics, such as emerging safety issues and new products.[18]

As it had with Bossert and Jani-Serv, Grainger eventually integrated Allied Safety into the core business.

Lab Safety Supply

Meantime, Grainger had completed another acquisition in safety supplies in 1992—the largest acquisition in its history—with its purchase of Lab Safety Supply for $160 million. Unlike the other specialty companies Grainger had acquired, this one would survive as a separate business unit. Located in Janesville, Wisconsin, Lab Safety Supply was the leading industrial and commercial business-to-business direct marketer of safety supplies in the United States, distributing such products as hazard control equipment, signs, personal protection equipment, and training information. The acquisition established Grainger as the preeminent safety products distributor in America and launched the company's entry into direct marketing.

Literally founded through the kitchen table discussions of Don and Gerry Hedberg in 1974, Lab Safety Supply quickly became known for its progressive work environment and innovative approaches to merchandising and customer service. Its first black-and-white catalog, in 1977, was 44 pages. In 1986 the catalog became full color. Two years later, Lab Safety launched its Safety TechLine, enabling callers to speak with safety experts about regulations, compliance, and product applications. By the time Grainger acquired the firm, Lab Safety had moved into a new headquarters building in Janesville and had expanded into 480,000 square feet of warehouse space.

"Grainger and Lab Safety were a good fit," said Larry Loizzo, Lab Safety Supply's vice president of marketing at the time of the Grainger acquisition. "We were actually excited about the shift because now we'd see firsthand what Grainger was like and how they went to market. As you look at your competitors in the marketplace, you always wonder what they look like from the inside. It was pretty enlightening for us. We've had a real healthy exchange of information back and forth on our business model and Grainger's business model and how each of us goes to market, and it's been beneficial for both parties."[19] Loizzo went on to become Lab Safety Supply's president.

Lab Safety's *General Catalog* was educational as well as user friendly. In 1993 the American Catalog Awards named it the best industrial catalog in the United States for the fifth consecutive year.[20] Each year the catalog added more products and reached more customers. By the end of 1995 it featured 34,000 items and served 350,000 customers nationwide. That year, Lab Safety nearly doubled its Janesville facility by adding 450,000 square feet of warehouse space. In addition, it began installing a bar code warehouse management system that tracked delivery of orders and improved storage, shipping, and delivery.[21]

Parts Company of America

In 1989, Grainger had separated its parts business into a division called Parts Company of America (PCA), which would provide parts and service to commercial and industrial equipment vendors. Throughout that year, the new division added a larger inventory of parts, including some that supported products not offered by Grainger, and set up its own marketing—a first for Grainger.

"We built the business with the marketing organization," said James Tenzillo, who started PCA's marketing functions before becoming Grainger's vice president for marketing development and planning.[22]

By 1993, PCA supplied an inventory of 110,000 repair and replacement parts for more than 550 equipment manufacturers. Known as "the business that never closes," PCA took phone orders 24 hours a day, 365 days a year, promising same-day shipping on orders received before 9:00 P.M. central time. Phone sales agents provided expert advice along with parts information.

In 1991, PCA consolidated its five locations into a headquarters building in Northbrook, Illinois, and opened a second phone center in Waterloo, Iowa.

The dedicated men and women of Parts Company of America handled more than 6,800 service calls a day in 1995. Customers called PCA to determine what part they needed. Service representatives had immediate access to all necessary resources to respond to customers' needs.

PCA formed strategic alliances with restaurants, hotels, and schools to fill their needs, sometimes taking total responsibility for a customer's purchasing, stocking, and parts inventory management.[23]

By 1994, PCA phone sales agents could fax parts information directly from their computers to customers, allowing quicker response to customers' needs.[24]

Consistency of business processes and documentation is a strength for any MRO supplier. In 1995 PCA's practices were rewarded with the ISO 9002 quality certification from the International Organization for Standardization. The group awards certification only when companies meet specific and stringent quality standards. PCA's certification was renewed the following year.[25]

In June 1995, the replacement parts business was renamed Grainger Parts. A new marketing approach repositioned the business so customers would recognize they could buy replacement parts from the same company where they purchased their MRO products.

Pleasing Everyone

By 1991, the national and international economies appeared uncertain. Many companies would shut down or make drastic staff reductions, but Grainger continued to grow, opening branches, gaining market share, and acquiring companies. As a result, Grainger was able to do a rare thing in the business world: please investors, employees, and customers alike.

Historically, investors in Grainger have rarely been disappointed in the company's performance, and that didn't change as Grainger continued its rapid growth. In 1991, the company declared a two-for-one stock split, paying dividends of 61 cents per share. By 1992, when W. W. Grainger, Inc., celebrated the 25th anniversary of its public offering, the stock had split four times since the original

$19 shares had been issued. One hundred shares of stock purchased on March 29, 1967, had become 1,600 shares. The company continued paying above-average dividends in 1993, 1994, and 1995.

Grainger employees continued to enjoy generous benefits. A survey revealed Grainger's benefits package to be 15 percent above the average package offered by 15 other major employers. In fact, Grainger's benefits package, which contributed 44 cents in benefits for each payroll dollar, outperformed the benefits package of the average American company, which contributed 37 to 38 cents for each payroll dollar.

"We do an analysis every year with a consulting company that ranks our benefits package with some of our larger competitors'," said Nancy Thurber, director of benefits. "They've concluded that our retirement package is far above the average, and that's including a company that might have a 401(k) pension plan. Our combined retirement package for just our profit sharing plan is way above the mark. Plus it's more than competitive in that employees don't contribute toward it; it's an employer-contributed program. So even though we have some newer employees who wonder why we don't have a 401(k), we ask them why they would want to defer their own money into a 401(k) for the company to match when our match on average is 15 percent to 16 percent of pay."[26]

By 1991, the cost to the company of providing its benefits package, including the health benefit plan and the profit sharing trust, exceeded $96 million. As the PST marked its 50th anniversary, what had begun as a modest company contribution of around $40,000 had grown into a $237 million fund.

Meantime, Grainger continued rewarding its workforce by adding yet more benefits. In September 1990, the company introduced the employee assistance program, a confidential counseling service available to all full-time employees, their family members, and part-timers referred by the company. The following year, Grainger welcomed back 21 employees who had served in the Gulf War. Those Gulf War veterans who had been with Grainger a minimum of 30 days and who planned to return to work within 90 days of their discharge received a special bonus check paying them for the period they had been on active duty, up to six months.

As a testament to good employee relations, Grainger has always been a nonunion company. "We don't need anybody helping us get along," said David Grainger. "If you get a union, it means your supervision has been bad."[27]

"You know, there's a reason why we have 16,000 employees and we're virtually union free," said Jere Fluno shortly before his retirement in 2000. "It's because we've treated our employees very, very well."[28]

Catalog Makeover

While its Specialty Distribution Group was getting under way, Grainger strove to improve its core business of providing MRO products to a growing number of companies.

For one thing, Grainger started the decade with a new look for its catalog. In January 1990, the *General Catalog* got a lot bigger. Rather than the 5.5-by-8.25-inch catalog that Grainger had been printing for more than half a century, issue No. 377 measured 7.25 inches by 9 inches and offered 48 percent more capacity for the company's expanding product line. Grainger Division continued to add thousands of high-quality items to its product line so that by the end of 1994, the *General Catalog* boasted 61,000 items.

Introduced in 1991, Grainger's internally developed Electronic Catalog on CD-ROM was designed to help the company's larger customers order with more speed and convenience. Customers typed in a product's description, brand name, or other keyword to receive product information, pictures, an order pad, material safety data sheets, and the location of the nearest Grainger branch. The CD-ROM also provided manufacturer model cross-references and a search index. In 1991, Grainger's marketing, field systems, and sales departments created a special sales package to promote the Electronic Catalog to the company's top 500 accounts. By year's end more than a hundred customers had ordered the Electronic Catalog.

"When the CD-ROM first came out, everyone was thinking of it in terms of benefiting the customer, and that turned out to be very true," said Jere Fluno. "However, the really big productivity improvement was at the branches. Visualize yourself on a telephone with a headset, talking to a customer and working with that paper catalog versus having it at your fingertips with a picture."²⁹

A Toe in the Water

Grainger's branch network was expanding as well. The company added four new branches in 1993, including its first investment in a local facility outside the United States, a 42,000-square-foot branch in Puerto Rico. "We were looking for credibility outside the United States," remembered Ben Randazzo, vice president for Latin America. "I felt that the safest spot or the best toe in the water would be Puerto Rico. And it did give us some exposure to dealing with customers outside the United States. It posed some logistical issues and some pricing issues, but we overcame those. Dick Keyser's support meant an awful lot."³⁰

Keyser admitted that Grainger's international activity had been "a very slow process." But he added that "the branch in Puerto Rico was a big step, and it has been one of the most successful branches we ever put in."³¹ Six months after the branch opened, Grainger's sales in Puerto Rico had increased 60 percent, and in only three years, sales had surged to the level originally projected for the eighth to tenth year. Grainger's entry into Puerto Rico was so successful, in fact, that in July 1998 it opened a second Puerto Rican branch in Mayaguez. That branch quickly gained credibility by providing customers with generators and other emergency supplies after a hurricane struck.

In the Zone

Also in the first part of the decade, Grainger rolled out a new logistics strategy utilizing a network of zone distribution centers. The strategy enabled Grainger to respond better to customers' diverse needs by creating greater scale in shipping and more quickly replenishing local branches. Moreover, the zone distribution centers were able to handle complex orders efficiently by combining multiple products into a single order and a single shipment.

Each zone distribution center measured about 200,000 square feet, or approximately 10 times the size of most branches. In 1993, Grainger's first such center opened in Ontario, California, near Los Angeles, handling shipping transactions and daily replenishment of inventory for 23 branches. The following year, Grainger opened two more zone distribution centers, one in Atlanta and another in the Dallas–Fort Worth area. And in 1995, zone distribution centers opened in Chicago, Cleveland, and Cranbury, New Jersey.

"We started to migrate from a model where every branch did everything," Dick Keyser explained. "We were sort of the 'jack of all trades, master of none' in every branch, and we started constructing zone distribution centers in order to fix that—to move the shipping into larger facilities where we had more scale and to then focus the branches more on their local market."³²

Grainger continued adding new branches every year so that by the end of 1995 it operated 344 branches nationwide and in Puerto Rico.

The company improved its management of inventory with sophisticated computer systems that were able to forecast local demand, ensuring that branches always had quick access to products. At the same time, it improved product

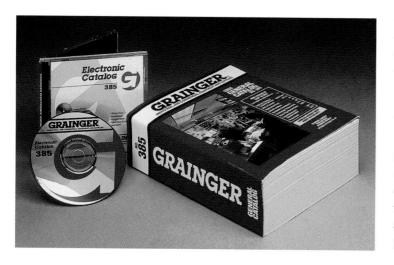

In 1991, Grainger introduced its larger customers to a high-technology alternative to the traditional *General Catalog:* the internally developed Electronic Catalog on CD-ROM. Customers who use the Electronic Catalog simply type in a description, brand name, or even industry slang for a product to receive product information, pictures, an order pad, and material safety data sheets.

The zone distribution center, which brings products closer to customers, is yet another innovation that Grainger created to satisfy its customers. In 1993, Grainger opened its first zone distribution center, in Ontario, southern California.

availability while reducing inventory with a system that monitored product movement from the supplier to the regional distribution center, the local branch, and the customer.

To Better Serve

A key part of Grainger's growth strategy was excellent service to all customers. In 1990 the company sped up sales transactions by replacing phone lines with transmittal of information via a satellite network that linked its 300-plus branches and three regional distribution centers.[33] Grainger, in fact, was one of the first companies in the nation to switch from a landline telephone network to a satellite network. The landline network was proving to be unreliable. Since much of Grainger's business depended on communication between the branches and distribution centers, it was a wise and timely move to invest in satellite.[34]

Ongoing training also has been an important factor in Grainger's excellent customer service. In 1990 the company established a College of Customer Contact. The 7,600-square-foot training center in Chicago included classrooms with computerized work stations, a simulated branch environment, and a product training area. New employees received three to six months' worth of experience in a one-week session at the training center. A similar center opened in Los Angeles the following year.[35]

Both training centers provided advanced programs for salespeople and field managers, and the Grainger Exceptional Manager Series enriched the skills of branch managers.³⁶

Grainger also formed Grainger Consulting Services to advise larger customers on reducing the overall cost of acquiring and storing MRO products. Consulting Services also helped customers overhaul their internal purchase process to reap the benefits of an integrated supply arrangement.³⁷

A profile of Grainger in the July 1995 issue of *Sales & Marketing Management* praised the company for its unique consultative selling approach and its training program:

> *The "quick hit" isn't in the vocabulary of Grainger's 1,400 salespeople.*
>
> *What is in their vocabulary? Do anything for a customer. Solve any problem. Establish needs before proposing solutions. These are the tenets drilled into Grainger salespeople's heads as soon as they're hired, and are constantly reinforced. "Once we get customers, we very rarely lose them because of our service package," says John Rozwat, vice president of sales.*
>
> *...The key to maintaining this doctrine? Continuous training.*³⁸

By 1994, Grainger's distribution centers were once again tight on space. Rather than build more warehouses, the company decided to reorganize existing space while improving inventory flow and customer service. "To better understand our options, we chartered a new, cross-functional team and called it the supply chain management team," said John Slayton, who became senior vice president for supply chain management. "The team included people from warehouse operations, transportation, logistics, inventory management, product management, marketing, sales, finance, and our branches." The team's goals were to increase customer satisfaction by improving product availability without driving up inventory and to increase warehouse capacity without investing additional money.³⁹

Thus in 1995, while Grainger was moving its headquarters to Lincolnshire, Illinois, the distribution center in Niles was transformed to warehouse the company's slower-moving items. The facility became known as Grainger's National Distribution Center. The change freed space in the regional distribution centers and branches, allowing better, faster service.

"That change has allowed us to make significant strides in our asset management," said Robert Pappano, retired vice president of financial reporting. "We've improved service level while taking inventory out of the system, and the analyst community is very impressed with that."⁴⁰

Supplier Relations

Grainger's efforts at excellence have extended to its suppliers as well, for the company takes pride in treating its suppliers as partners. "Our suppliers are an absolutely essential element of the value package that we take to the marketplace," said John Slayton. "If we don't have our suppliers and we don't have the products, what we have to offer in that value package becomes less and less."⁴¹

Or as Jim Baisley pointed out, "Every distributor, no matter how big it is, lives at the mercy of the supplier."⁴²

Grainger understood that but went one step further. Bill Grainger had discerned that a deal with a supplier wasn't good unless it was good for both Grainger and the supplier. That tenet stayed with the company long after Bill Grainger passed the reins of leadership to his son.

Cleveland Sales had been a Grainger supplier since Bill Grainger sent out his first postcard mailer, and that partnership continued into the next century through Cleveland Sales' manufacturing division, Clesco Manufacturing. Bill Grainger and a man named Fred Koehler

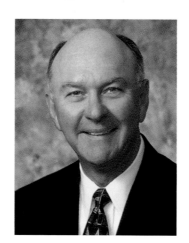

John Slayton, senior vice president for supply chain management, began at Grainger in 1979 as project engineer for the automatic storage and retrieval system.

met at the Sears Roebuck purchasing office in Chicago while they were each trying to make a sale. At the time, Bill Grainger was still working for Wagner Electric, and Koehler worked for an industrial brush manufacturer in Cleveland. The two men struck up a conversation, and Koehler suggested that Grainger could make a few extra dollars by selling accessories to go with the motors. It wasn't long before Koehler, with W. W. Grainger, Inc., as his main distributor, founded Cleveland Sales, which manufactured motor accessories such as motor shaft arbors.

"After Mr. Grainger opened up his motor store in Chicago, he proposed to Mr. Koehler that they put together a kind of kit to sell attachment accessories for the motors," said Joseph Malak, who joined Cleveland Sales during the Great Depression as a cleanup boy and eventually became its president. "It's been a good relationship ever since."[43]

As testament to the strong partnership between the two companies, in 2002, 30 percent of Cleveland Sales' revenue was with Grainger, according to Joseph's son, Thomas Malak, who became president after his father. "We've been dedicated to Grainger since our founding," he said. "We primarily have been a private labeler to Grainger, selling some products, like Dayton-brand products, on an exclusive basis. We consider ourselves like a subcontractor."[44]

That strong relationship with a supplier was Grainger's first, but it would not be the last.

"Bill and David Grainger believed that when times got tough and supply fell short, Grainger's suppliers would come to our defense because we had always treated them with the greatest degree of respect and integrity," said Don Bielinski.[45]

Experience supported that belief. Jere Fluno recalled an incident during the mid-1970s recession when a company that was not normally a Grainger customer wanted to buy rotary tillers from Grainger. "Why?" Fluno asked. "Because their supplier wouldn't sell them any rotary tillers. They had their whole production line coming to us. We had the same supplier, but we got the best service because we took care of the supplier. We paid the bills on time."[46]

For Grainger, having good relations with suppliers meant more than just paying the bills. It meant working with suppliers to improve quality and marketing of products. "Our product teams work with the suppliers and track what the return rates are for each and every product that we sell," said Timothy Ferrarell, senior vice president, enterprise systems. "When something gets out of whack, they're able to bring that up with the supplier. Grainger has had a history of making innovations with suppliers that make products more effective, lower their cost, and make them easier to install." As an example, Ferrarell noted that Grainger had been able to convince suppliers to package product in smaller quantities rather than let the distributor break it down and repackage it for the customer, which might decrease quality and cause a higher return rate. "We tell them, 'Let's make sure that we package it right so that it's efficient, so the customers don't have to buy 50 of them when they typically use them only one at a time.'"[47]

Fred Loepp, vice president for product management, noted that Grainger's high standards oftentimes motivated suppliers to improve their own performance. "The demands that we put on our suppliers tend to be much greater than what they get from anybody else," he said. "We measure their performance and work with them on how to improve. A lot of basic manufacturing companies are doing that with their raw material suppliers, but it's virtually unheard of in the distribution business."[48]

More than a few long-time suppliers agreed that working with Grainger was beneficial for both companies. "Grainger simply understands what a partnership with suppliers really means," said Michael Tellor, president of Rust-Oleum. "Grainger has pioneered a culture in which we work closely together to serve the needs of our common end-use customers, and keeping this focus creates a win-win situation for Grainger and Rust-Oleum. The highest level of mutual respect develops and is maintained."[49]

Tracy MacMillan, vice president of communications at North Safety, agreed that working with Grainger was a mutually beneficial relationship. As of 2001, North Safety had been supplying Grainger with personal protection equipment for more than 30 years and was, in fact, one of Grainger's first suppliers of safety products. "They've challenged us to support them with better processes and service levels and cost efficiencies," MacMillan said. "In doing that, we've enhanced our own service model. And in the same regard, we've challenged them to expand their safety program and to commit to a broader offering and dedicate sales and marketing time specifically for that product line."[50]

Likewise, Jim Lindemann, CEO of Emerson Motor Company, which manufactured the Dayton brand of motors, said that Emerson and Grainger had "an outstanding relationship that has been built on trust and performance."[51]

Fritz Zeck, president of Cooper Lighting, said that his company had chosen Grainger as an alternative channel to reach customers that Cooper hadn't previously been able to reach. "Our strategic partnership has paid dividends for both companies," he said.[52]

Brady Corporation's relationship with Grainger has grown progressively stronger and more fruitful, according to Mark MacDonald, business unit manager at Brady. "We have steadily increased sales in

Grainger's branches employ knowledgeable counter salespeople to help customers with their often unplanned MRO purchases.

the 13 years we've worked with Grainger," he said. "We each look at our individual company's goals and try to integrate them so that we're both going after the same thing."[53]

Recognition

Grainger was often rewarded by suppliers during the 1990s. At a ceremony held in London on May 14, 1994, Grainger was chosen from an international field of 30,000 suppliers as one of 171 winners of General Motors' Quality, Service, Price Award. The company was given the title in Grainger's category of "Worldwide Supplier of the Year 1993." By 2001, Grainger had received this recognition on six separate occasions.

On June 22, 1994, at a ceremony in Washington, D.C., the U.S. Postal Service presented Grainger with its Top Quality Supplier Award, selecting the company as one of 12 out of 60,000 postal service suppliers. And in 1995, Grainger won Abbott Laboratories' Corporate Preferred Supplier Award; Landis & Gyr's Excellence in Distribution, Partner in Business Solutions Award; the Supplier of the Year Award from Honeywell's Micro Switch division; and Milliken and Company's Distinguished Supplier Award.[54]

While Grainger continually improved customer service, it also decided to put greater emphasis on marketing. "Traditionally, Grainger had done precious little in terms of image advertising and those kinds of things," said Slayton. "We had a very large network of resellers throughout the country that were really doing the face-to-face work with a lot of our end users, and we were kind of in the back room."[55]

In the early 1990s, much of the public was not familiar with Grainger. Even Dick Keyser admitted that when he first considered a job at Grainger, he didn't know what kind of business Grainger was in.

That was about to change. In 1990 the company hired John Schweig, a strategic consultant for Grainger since 1986, as vice president for marketing. James Tenzillo, who had done so much in marketing for PCA, became institutional marketing manager. "We had to create a marketing organization where none existed," said Tenzillo. "Somebody was doing the marketing work, but most of it was in either product management or sales management."[56]

Meanwhile the sales department continued its own marketing efforts. In 1990 it began a new sales approach that established generalist and specialist sales representatives. Most Grainger customers would continue to be served by the typical Grainger sales representative, or generalist. But Grainger's large customers often had complex purchasing processes that required a salesperson with in-depth knowledge of their industries. Grainger's specialist sales force was developed to better serve these customers, which often represented national accounts and specialized industries such as health care, property management, hotel management, and education.[57] Grainger's sales force was so successful, in fact, that in 1994 it was awarded the H. R. Chally Group's World Class Sales Excellence Award. Grainger was one of only 10 companies in the country to receive the award.[58]

Meanwhile, Grainger took to the airwaves on ESPN in April 1993 with its first television commercial. The cable network had offered Grainger 10 commercial slots worth around $70,000 during broadcasts of "ESPN Outdoors," a morning fishing show, as a bonus to the company for running an ad in the outdoor magazine *Field & Stream*. The commercial began with a *General Catalog* rolling through the air in slow motion while line after line of products appeared in the background. A voice-over talked about products that "keep a business running smooth." On cue, the Grainger name appeared on the screen.

"The name Grainger was hardly advertised until well after I came here," said Dick Keyser. "Back then the whole advertising budget went into the catalog—that was it—and the only visibility was from a couple of our private label brands, Dayton and Teel. They never promoted or advertised the name of the company. There'd just be a little bit of trade advertising, so it's not surprising that we had a low profile."[59]

Outsourcing

The company for years had worked to be the supplier of choice by giving customers access to tens of thousands of preferred products, plus important information to help them choose products

Opposite: Grainger offers same-day shipping service from each of its branches and employs its own fleet of trucks to deliver products in selected major metropolitan areas, such as Chicago.

that would work best for them. Furthermore, Grainger had perfected its logistics system through its zone distribution centers so that even the most complex transactions were handled smoothly. The company also provided expert consulting that helped customers work out any kinks in their procurement process.

In the mid-1990s, Grainger took the next step that positioned it to be the integrated supplier of choice. The company began forming strategic partnerships with "best-in-class" distributors of products. Though these distributors sold their products through Grainger, they continued servicing their customers by providing technical assistance. Grainger also formed strategic alliances with companies in which Grainger and the alliance partner became preferred suppliers for each other when a customer asked either of them to take over some or all of its MRO procurement.

In 1994, Grainger formed an alliance partnership with Kennametal, a manufacturer and distributor of high-quality carbine cutting tools. Grainger became Kennametal's only national distributor. That year, the company also formed alliance partnerships for Norton Company's Abrasives Marketing Group; Square D's electrical control and distribution products; Bussman's fuses; and General Electric's lighting products.[60] The following year, Grainger added more companies to its best-in-class distributors: Cadillac Plastic and Chemical Company, which distributed plastic sheet, rods, tube, and film; and Crescent Electric Supply Company, which distributed electrical supplies and offered support services. Grainger also formed alliance partnerships with Motion Industries, a distributor of bearings and mechanical, electrical, and fluid power equipment, and Ferguson Enterprises, a distributor of pipes and valves.[61]

As more companies realized how costly procurement of MRO supplies could be—whether because of buying inefficiently, storing too much inventory, or unnecessarily using too many different brands—they sought to lower costs by outsourcing to integrated MRO distributors such as Grainger. "The whole business world is trying to get down to fewer and fewer sources," explained David Grainger.[62]

Smaller companies' MRO procurement, however, did not involve complex purchasing processes. Grainger was able to help them lower costs by identifying and finding products and making it easy to place their orders. This was a service Grainger had specialized in for many years through its *General Catalog* and branch network, and the service was about to get even better with the introduction of Grainger.com, the company's first Web site.

In 1999, Richard Keyser, David Grainger, and Jere Fluno shared in cutting the ribbon that opened Grainger's new headquarters in Lake Forest, Illinois.

CHAPTER ELEVEN
A COMPETITIVE ADVANTAGE
1996–2002

DURING THE MID-1980S, ANAlysts had estimated that the U.S. wholesale market was generating $70 billion–$90 billion a year. At that time, Grainger held a mere 2 percent of the highly fragmented market, though its share was larger by far than any of its competitors'. The company had grown significantly since then, but in 1999, analysts estimated that Grainger still held only 2 percent of the MRO market, which by then exceeded $250 billion.[1] In essence, Grainger's position had hardly changed in over a decade. However, Richard Keyser and Grainger's other leaders weren't alarmed. Rather, they saw the 98 percent of the market the company didn't have as a vast opportunity for expansion.

And expand Grainger did—into realms that no one had dreamed would exist when Bill Grainger founded the company on the eve of the Great Depression. With 1996 sales of $3.5 billion and earnings of $208.5 million, Grainger was well positioned to carve out a larger piece of the MRO pie. Throughout the latter half of the decade, Grainger became more and more agile in the increasingly competitive marketplace. The company not only embraced burgeoning technologies but expanded its bricks-and-mortar network and reorganized and fine-tuned its processes and procedures—all to improve its customer service.

Spurred by competition and a burning desire to serve all of its customers' needs, Grainger continually reinvented itself, combining products, services, and information to give businesses and institutions what they wanted, when they needed it, at a lower cost than if they went through other distribution channels. Grainger, in fact, emerged from the 1990s as North America's leading business-to-business provider of MRO supplies and related information through both Grainger.com and the more traditional bricks-and-mortar network.

Grainger.com

Grainger's electronic commerce activity had begun when it first issued its catalog on CD-ROM in 1991. But after "beating the data into submission to force it onto a CD," according to Barbara Chilson, who spearheaded Grainger's early electronic initiatives, the designers were surprised when they discovered that customers wanted real-time price and availability.[2] The company debuted its Web site, Grainger.com, in 1995, but it wasn't until

---

A sculptured plaque honoring William Wallace Grainger, the company's founder, is located just inside the main entrance of Grainger's new headquarters in Lake Forest, Illinois.

an engineer in Denmark saw the site and sent an e-mail saying he wanted to order fuses that Grainger realized it needed a way for customers to order on-line. By 1996, customers could access Grainger.com to search Grainger's entire product line (with more than twice the number of products in its paper catalog) and place an order on their regular account via the Internet. Ordering on-line would be commonplace in just a few years, but when Grainger began offering the service, it was way ahead of the game, especially in the MRO field, and many people wondered about the company's tactics.

Eventually the skeptics saw the light, realizing that customers could use Grainger.com's search capabilities to find any of thousands of products, along with quick access to detailed product information and on-line ordering. Not only was Grainger.com a convenient alternative to the more traditional catalog shopping; it also reduced procurement costs associated with administrative time and paperwork.[3]

Grainger felt that as an early adopter of business-to-business Internet commerce, it had both an opportunity and a responsibility to help customers become more familiar with the Internet's capabilities and more comfortable with its use—and it took advantage of those opportunities. In a very short time, Grainger.com had become the largest one-stop Internet catalog of MRO supplies. In 1997 and 1998 the site was rated one of the 10 best business-to-business Web sites by *Advertising Age's Business Marketing* magazine.

"At Grainger there is a spirit of experimentation," said George Rimnac, vice president and chief technologist. "The MRO industry is unexplored territory. It's a cranky, old-fashioned business, and in the middle is a company called Grainger that is very technologically advanced and is doing some very exciting things. The people who are involved in information technology are capable of making a real difference in the success of the business."[4]

During the next several years, Grainger.com introduced several new features, such as MotorMatch, which let shoppers key in specs for motors and find ones that matched; a feature that let special-pricing customers see their pricing on-line; and a feature that alerted customers whether a particular branch had a product in stock.[5]

All of Grainger's Web-site enhancements were designed to meet specific demands of customers. Grainger.com took customer service one step further by displaying information about preparing for emergencies such as hurricanes, tornadoes, floods, and earthquakes. The information helped businesses and individuals prepare for disasters before they occurred and assisted in protecting businesses and their employees.[6]

As the months passed, the company gained more experience in the fast-developing Internet technology while expanding the site's product offerings and fine-tuning its search and purchasing capabilities. At the same time, Grainger.com was earning more kudos from industry experts. An August 1998 issue of *Industrial Distribution* magazine pointed out how much Grainger's competitors could learn from its Web site.[7] *CIO* (a publication for chief information officers) named Grainger a winner of its 11th Annual CIO-100 Awards competition for reinventing the way it did business to take advantage of emerging technology.[8]

In the fall of 2000, Grainger upgraded and redesigned Grainger.com to improve customer satisfaction, adding new features such as faster order processing, customer-driven content, personal

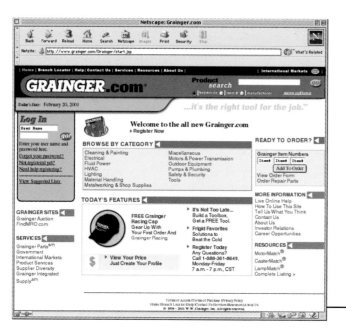

Revamped in the fall of 2000, the "all new Grainger.com" offered many added features that made ordering from Grainger more convenient than ever.

shopping lists, and enhanced graphics.[9] That year Grainger.com's sales increased 164 percent over 1999's to $267 million.

More improvements came in 2001, when Grainger introduced other advanced search options to the site. The national stocking number (NSN) special refinement let government customers locate government-approved products. Grainger was the only MRO distributor to offer NSN searches without first going through a set-up process. Also, the Web site's supplier diversity refinement allowed customers to quickly find products from Grainger's diversity suppliers, thus allowing them to more effectively meet their company's diversity purchasing requirements.

Focused Business Units

Over the years, Grainger had perfected the business of distributing to the local market. The company grew by adding more products and sales representatives and by covering more geography. Much of Grainger's success, in fact, could be attributed to its local availability and the high level of convenience it offered.

In more recent years, however, Grainger branched out and began offering a variety of ways for customers to access its products and services, for example, through its national accounts program and Grainger.com. In the latter half of the decade, the company expanded its footprint by giving customers even more choices. After all, each of the more than 1.3 million businesses Grainger served in 1996 had unique needs based on size, purchase situations, and preferences. Moreover, those needs frequently changed as different situations arose.

Thus in 1997 Grainger divided the corporation into focused business units, each targeting a unique market to provide optimal convenience and customized, low-cost MRO solutions. Under the reorganization, Wesley Clark, who had become senior vice president for operations and quality, was named group president with responsibility for Grainger Industrial Supply and Grainger Parts. Donald Bielinski became group president with responsibility for Grainger Integrated Supply, Grainger Consulting Services, Lab Safety Supply, and the company's various Internet

Grainger Industrial Supply is the nation's leading broad-line MRO supplier, providing customers with rapid access to 600,000 MRO products through a network of 390 branches.

businesses, including Grainger.com. A new unit, Grainger Custom Solutions, originally reported to Don Bielinski, but in 2000 it was transferred to Wes Clark. Before the reorganization, Grainger had formed separate business units to focus on its operations in Mexico (part of Bielinski's group) and Canada (part of Clark's group).

Though each business unit was tailored to meet a specific customer segment's needs, Grainger remained a single-minded organization under the overall leadership of Richard Keyser. In 1997 Keyser added the responsibility of chairman of the board to his duties as CEO. David Grainger, meanwhile, remained involved in the company as senior chairman of the board, though he admitted that, as his father had done, he was "letting go."[10]

Grainger Industrial Supply

Grainger's traditional domestic, branch-based network became known as Grainger Industrial Supply, a business unit that helped solve customers' immediate MRO needs, most of which were unplanned purchases. If, for example, an assembly line shut down for want of a new motor, the factory plant could turn to Grainger for a quick and reliable solution.

Grainger's small and medium-sized clients—businesses that represented more than 1.1 million of its customers in 1996—were able to conveniently reduce their total cost of MRO supplies by ordering through Industrial Supply's easy-to-use General Catalog, CD-ROM, or Grainger.com Web site or by picking up supplies from a nearby branch. Larger customers also used Grainger Industrial Supply to solve their immediate MRO needs.

With each passing year, Grainger Industrial Supply added new branches, ensuring that customers would always have access to needed products. The company added a slew of new products each year and continued to offer them for immediate pickup, same-day shipment, or delivery. Some of the new products Grainger distributed were value-priced private labels such as Westward tools and LumaPro lighting. Based on the popularity of such items, in 1998 the company created Grainger Global Sourcing to acquire high-quality products from other countries and sell them at competitive prices within the United States.

By the end of 2000, Grainger Industrial Supply offered rapid access to 600,000 MRO products through its network of 373 branches throughout the United States and Puerto Rico. Its trademark catalog had swelled to a 7.5-pound behemoth of 4,000 pages containing 85,200 catalog items; it had a circulation of just under 2 million.

To better manage inventory, in 1996 Grainger established its Smart-LINQ system to reduce overstock inventory at company locations. Moreover, the company developed a system that helped branches stock products for local requirements, thus reducing unneeded inventory while providing maximum service to local customers. The result was an inventory distribution system that was the envy of the competition. "We've got some of our best minds in our inventory management group," said Rick Adams, who was promoted to vice president of supply chain development in December 2001. "We have a world-class inventory management system, and the systems are homegrown."[11]

In 20 years, Grainger's catalog evolved substantially. The spring 1981 issue offered 9,100 products, the 1997 issue featured more than 78,000, and the 2002 catalog had around 100,000.

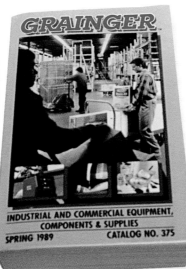
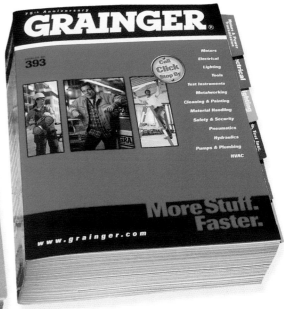

Though smaller than the older models, the new Grainger branches, like this one in Quincy, Illinois, featured open layouts and new ways to merchandise products. As an added benefit, the new branch design cost less to build and about half as much to operate compared to the older models.

At the same time, Grainger upgraded its information systems with state-of-the-art software that let the company better manage its network. And Grainger continually trained employees to provide optimum customer service.[12]

Gradually, as more customers used the Internet to order supplies, Grainger's zone distribution centers took on more shipping transactions, while the local branches concentrated on will-call and walk-in business. In 1997, the zone distribution center in New Jersey introduced a new operating process that let it support double the number of branches.[13]

Meanwhile, Grainger Parts continued to sell replacement parts and accessories. Customers had access to 285,000 parts by phone. In-stock orders placed by 9:00 P.M. Monday through Friday were shipped the same day from the distribution center in Northbrook, Illinois. Grainger Parts had trained customer-service agents who used detailed parts diagrams to help customers identify which part they needed and how to install it.[14]

The changes at Grainger were a result of the company's renewed focus on the customer's need. "We're making a shift that is somewhat subtle, but I think huge," said Wes Clark in a 1999 interview. "As opposed to being viewed as a service network, which feels somewhat passive—the customer has to initiate a need and then we do our best to service it—we're becoming more market driven and customer driven."[15]

Part of the shift Clark was referring to involved enhancements at the branches. The redesigned branches had wider aisles and eye-catching displays, contributing to increased sales and heightened customer satisfaction. Clark explained how the branches were being remodeled and remerchandised.

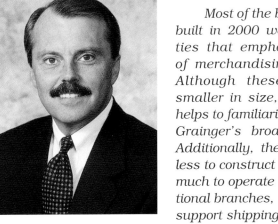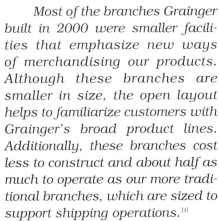

Most of the branches Grainger built in 2000 were smaller facilities that emphasize new ways of merchandising our products. Although these branches are smaller in size, the open layout helps to familiarize customers with Grainger's broad product lines. Additionally, these branches cost less to construct and about half as much to operate as our more traditional branches, which are sized to support shipping operations.[16]

In 1997, both Wesley Clark (pictured) and Donald Bielinski were named group presidents. Clark was responsible for Grainger Industrial Supply, Grainger Parts, and Acklands - Grainger Inc. He later became president and chief operating officer of W. W. Grainger, Inc.

At the end of 2001, Grainger celebrated the opening of its new distribution center in Los Angeles, the first of several in the improved distribution network. When completed, the network would be made up of five new centers and four redesigned ones.

In addition, Grainger changed its method for choosing locations for branches. "We used to spend the majority of our dollars on the facility and a much lower percent on the location," said Clark. "We would often be located in the back of an industrial park, thinking that if we had great availability and great service, our customers, being pros, would find us anyway. Now we spend a higher percentage on the location. We're trying to buy property on the major thoroughfares, where people will see us, or at the crossings or at places where our signage can be seen from a major highway."[17]

Grainger improved its logistics as well. Beginning in 2000, the company planned to invest more than $200 million to streamline and increase efficiencies in its distribution network. Rather than have local branches do everything from shipping orders to waiting on customers to answering phones, Grainger migrated to a different model: it planned to eliminate regional distribution centers altogether and grow the zone distribution centers.[18]

In all, Grainger planned to have nine strategically placed distribution centers across the country, which would allow it to ship 90 percent of orders within 24 hours. Keyser summed up the benefits of the new network, saying it will "enable us to remove a step from our distribution process, eliminate multiple handling of products, and get products to our customers faster."[19]

"We're betting our strategy on local availability, combined with a robust Internet front end and supported by an unparalleled logistics structure," Wes Clark told a group of analysts and business leaders at a conference in November 2000.

"We are working to align our organization and the Internet opportunity throughout all of our operations. And as we continue to bridge that gap between the Internet and traditional MRO, one thing is certain. We will remain customer focused."[20]

By 2001, the redesigned distribution and logistics network was well underway. New distribution centers opened in Dallas and Los Angeles, and ground had been broken in Jacksonville, Florida.

Lab Safety Supply

Lab Safety Supply was separate from Grainger Industrial Supply but still in the distribution business, the one business in the Specialty Distribution Group that remained an autonomous operation.

Lab Safety Supply offered safety and MRO products through the Internet and direct-mail catalogs and was geared to customers who did not require face-to-face interaction. "Our customer base runs the full gamut of what's out there," said Larry Loizzo, president of Lab Safety Supply. "We deal with a lot of different customers, from the top Fortune 500 companies to individuals who may be in a consulting capacity advising somebody else and buying products."[21]

Lab Safety Supply had enjoyed unprecedented success since Grainger bought it in 1992. In 1996, it expanded its warehouse by 195,000 square feet and established a self-designed Web site at Labsafety.com, which quickly became the largest safety resource on the Internet. The site included safety and product-related information, customer newsletters, trade show schedules, and an on-line catalog request form.[22] This important information was accompanied by free technical support offered to all customers through a toll-free hotline called Safety TechLine. The Safety TechLine call center assisted customers with everything from product specifications and chemical compatibility to regulatory compliance issues and general health and safety concerns.[23]

Also in 1996, Lab Safety Supply expanded its award-winning catalog's product line by 20 percent, mostly in material handling and maintenance products. That year Lab Safety Supply offered its first *Material Handling Direct* catalog, targeting customers who bought such supplies in conjunction with safety supplies.[24] The new catalog was so successful that the following year Lab Safety Supply launched *Maintenance Direct*, which featured more than 400 pages of facilities maintenance products. After that, Lab Safety Supply began offering targeted catalogs for a variety of product categories, including first aid, health, and wellness; labware; shop supplies; safety essentials; signs and labels; and spill cleanup.[25]

By 1999, Lab Safety Supply's main catalog featured more than 46,000 products, many of them non-safety-related, for which demand was quickly growing. In 2000, Lab Safety Supply produced 41 different catalogs and laid plans to expand into more targeted product categories as the government put more emphasis on worker safety.

Lab Safety was doing well enough (earning more than cost to capital) that the company decided the best way to grow the business was to focus on acquisitions and increased sales. In February 2001, Lab Safety Supply purchased the privately held Ben Meadows Company, based in Canton, Georgia. Ben Meadows, which had annual sales of $20 million, was a business-to-business direct marketer that specialized in equipment for the environmental and forestry management markets. The acquisition made Lab Safety Supply a leader in these markets, and Ben Meadows's broad customer base allowed Lab Safety to expand its market base.[26]

Under the leadership of Larry Loizzo (above), Lab Safety Supply became the leading North American business-to-business direct marketer of safety and other industrial products. By 1999, its award-winning catalog (left) featured more than 46,000 products.

Grainger Integrated Supply

Meanwhile, the company offered a variety of services to help larger customers manage their more complex MRO purchasing. A study had revealed that for every dollar spent on MRO supplies, 60 cents went toward the actual cost of the product, while 40 cents went toward the process of getting the product. "There's very little the economy can do to dramatically change that 60-cent component of the product cost because that's really the manufacturer cost and the efficiency of the logistics channel to get that product to the end user," explained Bielinski. "But there's a lot that can be done about the process-cost component, the 40 cents."[27]

Grainger Integrated Supply sought to reduce that 40-cent process cost by taking over the management of customers' MRO procurement, thus letting them focus on running the business itself. Those customers who chose an integrated approach were willing to outsource their entire MRO materials management and, if necessary, totally revamp their processes on Grainger's advice.

"The integrated supply business is unique because we're not in the distribution business," said Peter Torrenti, former president of Grainger Integrated Supply. "We're a management services company. We provide good, proactive, strong, indirect materials management, and we do it on-site at our clients' place of business. We create value in the services that we render, and we get paid by driving cost savings."[28]

As Grainger Integrated Supply offered additional services, more and more large companies turned to it for their MRO solutions—and it was no wonder. As an industry consultant with Frank Lynn & Associates in Chicago said, "For some companies, outsourcing MRO supplies and services can offer long-term savings of 3 to 15 percent, resulting in savings of up to $300,000 or more annually."[29]

Grainger Custom Solutions

Dealing with large customers' needs was, at first, somewhat of a challenge for Grainger because the company wasn't set up to fulfill such large orders. If, for example, Ford Motor Company wanted 1,000 of a certain item within 24 hours, Grainger would have to strip the stock in its branches to fill the order. "We don't have 1,000 of anything in any one branch," explained Wes Clark. "We might have two in every branch because, statistically, nobody ever wanted more than one. So we'd have to starve all the branches in order to get 1,000 for Ford. Meanwhile, we'd be out of stock for all the regular customers who come to our branches."[30]

Grainger Custom Solutions was formed for large companies that required materials management services but didn't want to outsource their procurement process or employ on-site management services. Businesses that contracted with Grainger Custom Solutions were able to reduce their total MRO costs by selecting a primary distributor for each of their major MRO product needs and letting Grainger provide a low-cost logistics network for the repetitive purchases.

Grainger Integrated Supply Operations (GISO) held special classes under the name GISO "U," where employees were trained to use Grainger's new enterprise resource planning (ERP) system.

"These large customers are looking for suppliers that can do a lot of things for them," explained Mike Kight, then president of Custom Solutions. "It allows their core business to focus on what they do well and allows us, on a lower-cost platform, to get sharper and better at what we offer this segment of customers."[31]

As Grainger realized that many of its Custom Solutions customers were still coming to the branches for unplanned purchases, it scaled the business unit back in early 2000.[32]

Grainger Consulting Services

Sometimes customers didn't know which approach would serve them best. For this group Grainger offered Grainger Consulting Services, which determined how customers could best reduce their total MRO costs based on individual needs. Understanding that the lowest total cost was more relevant than individual product price, Grainger Consulting Services helped customers manage their MRO supplies by determining the best process to lower the customer's materials management costs, then outlining the steps the company needed to take to reach its goal.

In addition, Grainger Consulting Services provided customers with the country's most comprehensive database of MRO products (more than 1.3 million items) and was widely recognized as the leading professional services firm specializing in MRO materials management consulting.[33]

Serving Mexico

After the passage of the North American Free Trade Agreement (NAFTA), which encouraged the fair exchange of goods and services among North American countries by lowering trade barriers, Grainger saw an opportunity to expand its North American operations. In April 1996 Grainger opened a branch in Monterrey, Nuevo León, Mexico.

While Grainger, S.A. de C.V., represented a new type of venture for the company, Grainger was by no means a stranger to the international marketplace. By 1996, 12 branches bordering Mexico and Canada were already exporting products to those countries. Four other branches were serving nations throughout the world. Grainger Caribe, Inc., in Puerto Rico, was exporting to the eastern Caribbean, Hispaniola, and the Virgin Islands. The Miami

Left: Mike Kight began his Grainger career in 1967 as a part-time branch worker. He graduated through various positions, serving as president of Custom Solutions and, more recently, president of Integrated Supply.

Below: Grainger's branch sales people are trained to help customers find exactly what they need—even when customers don't know the precise product they're looking for.

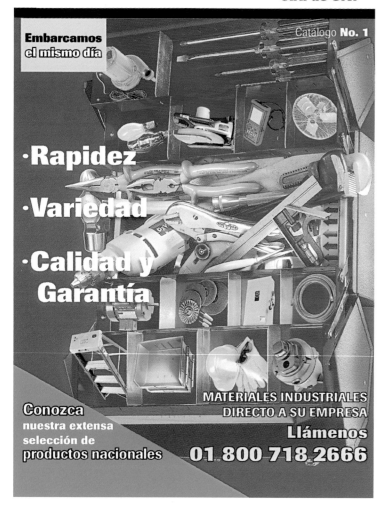

In the spring of 1996, Grainger opened its first international branch, in Monterrey, Nuevo León, Mexico. To better serve its Mexican customers, Grainger printed a Spanish-language version of its traditional big red catalog.

branch was exporting to the Bahamas, the Caribbean, the Cayman Islands, Central and South America, and Jamaica. The branch in Morton Grove, Illinois, was exporting to Africa, Bermuda, Europe, India, the Middle East, and the former Soviet Union. And the Oakland, California, branch was exporting to Asia and the Pacific.

Grainger, S.A. de C.V., positioned the company to better serve multinational customers and the Mexican market. "Mexico was kind of an untapped frontier," said Ben Randazzo, vice president for Grainger's Latin American operations. "We knew that an awful lot of our national accounts were moving down there, so the battle cry became, 'Hey, our national accounts need us.' "[34]

Getting started in Mexico was certainly more arduous than it had been in Puerto Rico due to the complexities of border logistics and regulations. When it began operations, the branch in Mexico offered 39,000 products from its 40,000-square-foot facility and was able to ship products to the industrial part of Mexico (which made up about 80 percent of the branch's business) within 48 hours. Randazzo noted that "speed and convenience have a different dimension than in the United States. In Mexico, speed and convenience mean getting product in two or three days because nobody else has inventory."[35]

Grainger's business in Mexico increased as more customers took advantage of the branch's wide array of rapidly available products and convenient ordering process. In 1997, Grainger, S.A. de C.V., began adding products from Mexican suppliers. The following year the branch doubled in size to 80,000 square feet. By the end of 1998, customers in Mexico had access to more than 60,000 products. It was little wonder that companies in Monterrey rated Grainger as their first choice for miscellaneous MRO products. In 2000, Grainger opened its second Mexican branch, in Guadalajara, this one much smaller, to gain a more local presence. By that time Grainger, S.A. de C.V., offered more than 70,000 broad-line MRO products. Additional branches have since opened in the Mexico City, Puebla, and Tijuana areas of Mexico.

Acklands - Grainger Inc.

In December 1996, eight months after it opened its first branch in Mexico, Grainger completed the acquisition of the Canadian industrial distribution business of Acklands Limited. The company was Canada's largest distributor of broad-line industrial supplies and served all the Canadian provinces. Wes Clark explained that there were two main reasons for the acquisition: "One was a defensive play," he said. "We didn't want to have a very large competitor stretched coast to coast north of us. There was also an offensive play, though. Acklands gave us a great opportunity to expand with a lot more purchasing power and a tremendous amount of geographic coverage."[36]

A carriage maker and blacksmith named Dudley Acklands had founded the Canadian company in 1889 to provide wagon parts and tools to others in his trade. In 1905 his son, J. D., joined the family business. From there the company expanded its operations by adding branch warehouses, increasing sales of industrial supplies, and making strategic acquisitions.[37] Acklands had a reputation for supplying literally everything from A to Z; it had actually supplied apples in the far north and a zamboni for a city in the Northwest Territories.

Douglas Cumming had been with Acklands since 1956 and retained his position as president of the division when Grainger acquired it. Throughout Acklands's history, Cumming said, it had been "a bit of a miniature version of Grainger. Our branches were smaller, and they were in smaller populated areas. We followed the smokestacks. If there was a gas plant, a pulp mill, or some other kind of industry, we would be there, and eventually we covered almost every major and minor town in western Canada."[38]

Keyser agreed with Cumming's assessment. "They were about as close to Grainger as you could find in Canada," he said. "In fact, they really were the Grainger of Canada. And so when the company came up for sale in the summer of '96, we sent a team of operations people up there to take a really close look. They came back and said, 'Hey, this is us 15 years ago.' It was like looking in the mirror in terms of branch appearance, the level of systems support, and the way in which they operated."[39]

The unit of Acklands that Grainger bought, renamed Acklands - Grainger Inc., had sales exceeding U.S.$300 million in 1995. With 178 industrial stores and seven distribution centers throughout Canada, the business greatly enhanced Grainger's already strong branch network in North America.[40]

"They appeared to be running pretty lean, pretty effectively," Keyser remembered. "There looked to be a lot of opportunity to grow and not have to go in and cut and slash the overlying business. The employees up there were absolutely delighted to be with somebody with long-term focus on business, and we were delighted to have them. I guess it's a true North American footprint."[41]

The two companies shared a lot of synergies and were "a beautiful cultural fit," according to Keyser. Wes Clark's efforts ensured that the integration of Acklands into Grainger went as smoothly

After Grainger purchased the Canadian industrial distribution business of Acklands Limited, it formed Acklands - Grainger Inc. in December 1996. By 2000 the new company had 188 branches throughout Canada and provided access to more than 300,000 MRO products.

THE GRAINGER NAME: POPULARIZING A BRAND

TRADITIONALLY, GRAINGER HAD BEEN a very modest company—a company of strong character and quiet confidence that had never really felt the need to toot its own horn. But that rather reserved attitude was changing as the company began realizing the benefits of spreading the Grainger brand name.

"When I joined [in 1986], I was constantly amazed at how private the company was," said Bob Gideon, who began his Grainger career in communications and later became employee programs coordinator. "I think one of the directors said one time, 'You know, we're getting a little too big to hide our light under the bushel.' But I think that modesty comes from the personalities that have shaped the organization. We don't want to go out on a limb with a statement. We want to make sure that it's a very conservative statement—that it's very accurate and well thought out—and we're not going to make any wild claims."[1]

As the company grew, it became more proactive and began initiating discussions with analysts rather than reacting to their reports. At the same time, it broadened its brand image through advertising and other marketing channels.

"We're trying to become more focused in our marketing programs to get our customers to buy in more product categories," said Dennis Jensen, who retired in 2001 as vice president for field operations. "We believe this will give us more business and greater retention because customers are going to think of Grainger more for all their miscellaneous stuff versus thinking of us as niche players in lighting or tools or whatever."[2]

Wes Clark told *Advertising Age* magazine in 1997 that Grainger had been spreading its print budget too thin across trade magazines for a broad range of industries. With the print media being so cluttered with advertisements, Grainger decided to test the television medium. "We think we have an opportunity to distance ourselves [from the competition]," he said.[3]

After research showed that purchasing agents and maintenance and repair supervisors were likely to watch football, Grainger chose *NFL Monday Night Football* on ABC for

Throughout the late 1990s, Grainger worked hard to get the Grainger name before the public's eye. When physically challenged athletes met in Atlanta for the 1996 Summer Olympics, for example, Grainger and Port-A-Cool teamed up to supply world-class cooling capabilities to competitors (left). Grainger also sponsored a NASCAR racing team captained by Greg Biffle (opposite).

its first network television commercial. The 30-second spot with the theme "More ways to keep your business running right" premiered on November 10, 1997, and asked viewers if they "Need stuff?"[4]

Other marketing channels weren't quite so far reaching, though they were still effective. In May 1998, a first-of-its-kind Grainger van went operational to support Boston's central artery/tunnel project (also known as the "Big Dig"), a multi-billion-dollar infrastructure project that would supersede Boston's elevated, six-lane, central artery highway system. The Grainger van, equipped with a mobile branch computer system, served as both a moving billboard and a minibranch, providing workers access to all the company's products and delivery service, too.

Beginning in 1998, Grainger increased its brand awareness even further by sponsoring a Roush Racing team in the NASCAR Craftsman Truck Series (NCTS), with Greg Biffle as captain of the Grainger Racing Team. Driving a red #50 Grainger Ford F-150, Biffle earned the title "Cintas Rookie of the Year" that season. After a record-breaking nine victories the following year, the Grainger Racing Team finished second in the overall standings. The 2000 season was even better for Greg Biffle and the team; they won the NCTS Championship.

GRAINGER Racing

as possible. "I think Wes handled the assimilation in just the right way," said Jim Baisley. "It could have been a different experience if not for Wes Clark's ability and the receptivity of the folks at Acklands - Grainger to working with Wes. The cost of integrating the two companies ended up being a lot less than we were projecting because they were able to work together so well, and you have to give Wes an awful lot of credit for that."[42]

Though Acklands - Grainger Inc. offered the continent-wide distribution capabilities of its parent company, it remained an autonomous organization because of differences in the Canadian and U.S. markets. After its first year of operation with Grainger, it opened a new, 193,000-square-foot office and distribution center in Edmonton, Alberta, which consolidated some of the Edmonton warehouses. Also in 1997, Acklands - Grainger Inc. added several new product lines, including motors and lighting, and standardized its product lineup throughout Canada. The Acklands - Grainger Inc. catalog, printed in both French and English, premiered that year, offering more than 260,000 products.[43]

Acklands - Grainger Inc. continued to grow despite a weak Canadian economy. In 1998 the company expanded its presence into Canada's industrial center by opening seven new branches in eastern Canada and another distribution center that served the growing demand for its private-label, value-priced products, the Westward and Profast'ners brands.

By 2000, Acklands - Grainger Inc.'s coast-to-coast network totaled 188 branches, all electronically linked to major distribution centers in Edmonton, Saskatoon, Winnipeg, and Toronto. Sales exceeded U.S.$399 million. For large customers, Acklands - Grainger Inc. offered supply-management solutions. And with more than 300,000 products, it offered the most comprehensive selection in Canada. In addition, customers had access to 220,000 repair parts through Grainger Parts, plus repair service for everything the company sold. Customers' product questions could be answered by Acklands - Grainger Inc.'s highly trained customer-service staff or in-house technical specialists.

Also in 2000, Acklands - Grainger Inc. set off on a strategic five-year plan that involved growing the business to exceed Can$1 billion in annual sales. To reach that goal, the company set out to expand its business to large customers—especially in eastern Canada—by offering new services and capabilities. Meanwhile, it focused on small and medium-size customers through enhancements to its branches. Acklands - Grainger Inc. also sought to develop its e-commerce abilities through Acklandsgrainger.com. The company took its first orders over the Internet in July 2000. It also enhanced its recruitment and retention programs and paid special attention to becoming more efficient, both financially and logistically.[44]

"We're on track for our five-year goal," said Douglas Harrison, who joined Acklands - Grainger Inc. as president in 1999 after the company enhanced its presence in eastern Canada by moving its headquarters to Toronto. Cumming remained chairman of the board. "I believe strongly in continuous development and continuous learning of an organization, and I believe a leader is only as good as the people who are part of the team. The people at Acklands - Grainger have done a tremendous job in pulling together and looking at the future and being truly excited about the future."[45]

International Endeavors

Business in Puerto Rico, Mexico, and Canada was faring quite well. But a Grainger study led Keyser to conclude that he should not expand business overseas, at least the bricks-and-mortar part, because of the high risk associated with building warehouses and supplying inventory in a less familiar marketplace where there was less product compatibility.

"It's understandable why Grainger never spent a lot of time outside the United States," said Y. C. Chen, who, prior to becoming vice president of supply chain services, was vice president of international

Opposite: W. W. Grainger Inc.'s rich history includes having only three chief executive officers in 75 years: William W. Grainger (center), CEO 1927–1967; David W. Grainger (right), CEO 1967–1995; and Richard L. Keyser (left), CEO since 1995.

Internet commerce. "Grainger doesn't have control over the design and manufacturing of the products it distributes, so we can't modify the products to sell them to foreign markets."[46]

"Fulfilling foreign demand is a lot more intricate than just building warehouses," added John Schweig, senior vice president for business development and international. "In many parts of the world, the products we sell are not the products people want to buy."[47]

But the same study that convinced Grainger not to build overseas revealed something else. "During the process of going through and analyzing the situation in the foreign market, we came across an opportunity that Grainger hadn't been taking advantage of," said Chen. "We realized that whenever the manufacturers resided in a foreign country, we could be closer to them and purchase from them directly and therefore gain a bigger portion of that margin as opposed to paying an importer to do that for us."[48] Thus in 1997 Grainger International set up an office in Hong Kong, run by Chen, to procure private-label, value-priced products from the Asia Pacific region, eventually procuring from Europe and Latin America as well.

How to Keep Good People

As always, Grainger continued to attract and retain an excellent workforce by offering numerous training and development programs. Discover Grainger, for example, was a CD-based orientation/assimilation process for new employees that let them begin their training even before their first day of work. In addition, programs like the Leadership Development Program and the Financial Development Program provided Grainger's future leaders with the skills necessary to keep a competitive advantage in years to come.

Grainger also gave employees the opportunity to move around the company as they developed new skills. A significant percentage of Grainger's workforce was comprised of people who had been with the company for a decade or more. For many, Grainger was the only employer they had ever known. For Robert Thrush's 30th anniversary with the company, his colleagues pulled out his original application, which asked what position he was applying for. "Anything available," he had written

soon after finishing high school, not knowing that Grainger did, in fact, encourage employees to aim toward anything that would serve the company. Grainger had a reputation for fostering growth by providing employees with many choices.

Since Grainger was founded, company leaders have endeavored to provide the best working environment for employees through incentives, benefits, professional training, and just plain pat-on-the-back appreciation. Organized into separate business units, Grainger was able to customize its employee reward and recognition programs for each business. In 2000, for example, Grainger began "career banding," which focused on the success of the company by rewarding and compensating employees for performance and their ability to develop skills.

Painting Our Future

Part of the satisfaction of working at Grainger could be attributed to the company's guiding principles, the values for which it stood. Throughout the organization, employees knew that Grainger promoted quality, integrity, and honesty at all levels. In 1996 Grainger formalized those company values for all: agility, empowerment and accountability, ethics and integrity, having fun, learning, and teamwork.[49]

In September 1996, 450 of Grainger's leaders attended a conference titled "Painting Our Future," in which each of the values was discussed to determine if it was truly guiding behavior. Meantime, an artist created a rendering of the six values from which posters and value cards were made.[50]

Grainger's employees did indeed embrace the company's values. As Paul Wallace, retired vice president for financial services, who had been with the company since 1973, pointed out, "The list of values capsulized what's been here all along."[51]

A Pleasure to Work With

Ethics and integrity were perhaps the values employees appreciated most. "The integrity of the people who work here is phenomenal," said John Howard, remembering his first impressions of the company when he joined as senior vice president and general counsel in January 2000. "Mr. Grainger has driven a lot of his personal values through the company, so it comes from a very value-based, ethical approach."[52]

Jere Fluno, who retired from his post as vice chairman in July 2000 after 31 years of service, acknowledged that he had "never spent a sleepless

In 1996, Grainger held a "Painting Our Future" seminar (opposite) so employees could discuss the company's values. Once the values had been deliberated, the artist B. Marvin (below) painted a picture representing them. The result (above) displayed Grainger's commitment to agility, empowerment and accountability, ethics and integrity, having fun, learning, and teamwork.

night worrying about this company having done something illegal, immoral, or unethical. It's a fabulous company with fabulous people," he said. "Grainger was always on the straight and narrow."[53]

"People who are close to the organization admire Grainger for its integrity," said Nancy Hobor, vice president for communications and investor relations. "Externally, the media may not see the company's integrity or Wall Street may not see it, but those people who deal with Grainger—customers, suppliers, employees—they see it and appreciate it."[54]

Grainger's suppliers were quick to point out the company's enduring integrity. Campbell, which consisted of Campbell Hausfeld, Wayne Water Systems, and Power Winch, had been working with Grainger for more than 30 years. The relationship

> ## THE RIGHT STUFF FROM HELPING HANDS
>
> GRAINGER HAS ALWAYS GIVEN back to the communities it serves. At no time was that more evident than after the September 11, 2001, terrorist attacks, when Grainger's people worked together to help those affected by the tragedy, their customers, each other, and the nation as a whole.
>
> Grainger fully supported relief efforts by keeping its branch on Varick Street in Manhattan operating around the clock for several weeks. In the days after the disaster, Grainger's comprehensive distribution network allowed the company to deploy hundreds of semitrailers containing emergency products to New York and Washington, D.C.
>
> The first supplies were delivered to the Manhattan disaster site within a half hour of the tragedy, thanks in large part to Dave McCartney, Northeast regional inventory manager, Mike Jaffe, Mid-Atlantic regional inventory manager, and hundreds of Grainger employees. Unable to take a plane, McCartney and Jaffe drove more than 1,100 miles to Washington, D.C., while they hosted conference calls via cellular phone to connect with Grainger's field forces and coordinate speedy deliveries.
>
> In addition, the company contributed $1 million in cash and emergency supplies (including $25,000 to the September 11 Fund and $1,500 to the Twin Towers Fund). Supplies included respirators, bottled water, cots, emergency lighting, batteries, chargers, and tools. The company also matched employee contributions through the Charitable Matching Gift Program.
>
> Grainger set up a command center to keep employees updated on its support of relief efforts and to help them deal with the tragedy, on both personal and professional levels. It also set up an Emergency Resources section on Grainger.com and mobilized employees from several East Coast locations to provide assistance at branches near "Ground Zero."

started when Campbell Hausfeld supplied Grainger with air compressors under Grainger's private-label Speedaire brand. Dick Heiman, Campbell's president, noted that the relationship between the two companies was "special because of the people at Grainger. Our relationship with Grainger is probably the most ethical relationship we have," he said. "Grainger is the only customer I can think of that's advised us if we've overshipped them. Any customer that we undership is very glad to tell us about it, but they're one of the few that I can remember that advises us if we overship."[55]

"Grainger stands as one of the best examples of a company with high ethics," said Joe Ramos, group president of Rubbermaid. "Grainger is well focused and very customer focused. They're a company that suppliers such as ourselves absolutely love to work with because of what they stand for. Going to market with Grainger is an absolute delight."[56]

As Nancy Hobor pointed out, Grainger's culture and philosophy had been built up over time. "That can largely be attributed to David Grainger and to the legacy that the Grainger family personifies," she said. "It's been ingrained over many, many years."[57]

As David Grainger moved further from the company's day-to-day operations, Dick Keyser continued the tradition of setting a good example from the top of the organization. So did the other Grainger veterans. "Our culture starts from the top and filters down," said Nancy Thurber. "We have a lot of people who have been around for a while. When a new person comes in and sees the length of service that some of the people have here and the reasons why they stayed as long as they have, it just kind of builds on itself."[58]

"Some of the very traditional values of the company pervade this organization," said Dick Quast.

> Response from Grainger employees was so overwhelming that the company had to turn away volunteers who wanted to help in Grainger's relief efforts. Twenty-seven specially trained Grainger employees (part of Grainger's "Ready When the Time Comes" program) spent 29 days assisting the American Red Cross in Chicago in answering phone calls, serving food to passengers stranded at O'Hare International and Midway Airports, and assisting blood drives. Hundreds of other employees gave blood or donated food, clothing, money, and time.
>
> Jim Ryan spoke with many Grainger employees at different locations who asked him to tell their colleagues how much they appreciated what their fellow employees were doing to help. Ryan, too, expressed his appreciation in a letter to employees:
>
> *I've found that there are many of you who are working very hard, but feel you're not working hard enough—that you're not able to contribute enough to the relief efforts. Let me say now that your efforts in serving the day-to-day needs of our customers are just as critical.... I'm pleased to hear from many of you that knowing what's going on in New York and Washington, D.C. has helped create a greater sense of community among Grainger employees throughout the nation.*
>
> Richard Keyser and Wes Clark had similar praise: "Our pride for what Grainger employees are doing to help the relief and rebuilding efforts continues to deepen following the September 11 tragedy."
>
> Perhaps most noteworthy, however, was the response from rescue workers who depended so much on the products Grainger delivered. An army reserve soldier who came to a Grainger branch in the Bronx for supplies told the employees how vital Grainger was and returned later to take their picture. In Richmond, Virginia, a man collecting donations for transport to the disaster sites found out one of his donors, Patricia McKerns, was a Grainger employee. "With tears in his eyes, he told all the people there about what Grainger was doing and that we had a branch open around the clock to serve the efforts," wrote McKerns. "As I walked to my car, those around me kept saying thank you and shaking my hand."

"There is still a loyalty effect here. It's built into the fabric of the company. It has never been run for quarter-to-quarter performance. The company sets very high standards for itself, but it has always focused on a longer-term horizon. We do understand the tradeoff and the balance between short and long term, so while we're a large public company and have been for more than 30 years, in a way it's almost been run like a private company, and it is the loyalty that ties us together. There is a lot more employee concern here than in most other places about the long-term success of the company. People take it a lot more seriously."[59]

As testament to Grainger's good employee relations, *Fortune* magazine's January 12, 1998, issue named Grainger one of "The 100 Best Companies to Work for in America." The company received the same honor in 1999. *Fortune* selected companies based on compensation, benefits, training and development, internal communications, and employee recognition and involvement.[60]

"Grainger's secret is simple," said Keyser, acknowledging the award. "We have great people, and we do everything we can to keep them. We work to provide an environment where employees are free to do what they feel is right without navigating layers of bureaucracy."[61]

Enterprising Problems

As technology accelerated the rate at which Grainger's business grew, the company discovered that the effectiveness of its information services was fast diminishing, especially with the Internet changing the entire business. "We were out of sync with the pace of change in business," said James Ryan, who at the time was vice president for information services. "A lot of the systems that were running the

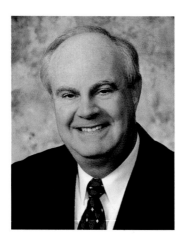

Left: Jere Fluno, who had helped Grainger go public back in 1967 and began working for Grainger in 1969, retired as vice chairman in 2000, with 31 years of service.

Below left: Grainger continued to prosper under the leadership of Richard Keyser, who became chief executive officer in 1995 and chairman in 1997.

company were as much as 15 to 20 years old. While having systems that lasted that long is a real testament to the quality of what we built, we recognized how important information technology was to the competitiveness of the company and began rebuilding the infrastructure to stay even more competitive."62

Rather than spend money maintaining its internally written information systems, which it would have to revamp for the year 2000, in 1997 Grainger decided to invest in a new enterprise resource planning (ERP) system. After careful consideration, Grainger chose a world-class software vendor to create a system that would replace the financial and operating software that ran the branches, distribution centers, and product management group.63

But the transition to the new ERP system, which debuted in 1999, was anything but smooth, and Grainger's 1999 sales and earnings paid the price. Sales increased by a mere 4 percent from the previous year, while the company's earnings dropped 24 percent. The disappointing year was mainly a result of unanticipated problems with the new system, which accrued expenses in installation and cost the company at least $20 million in sales.

"Any time you implement something of [the ERP's] magnitude, it becomes very complicated because everything is so new, and you're reliant on so many different vendors and parties," said Patrick Davidson, who became vice president for information services in January 2000, when Jim Ryan took over leadership of Grainger.com. Davidson was responsible for shoring up the reliability of the ERP system. "We had some hardware reliability problems and system configuration problems," he elaborated.64

Grainger also experienced problems transitioning to the complex system because employees had to be retrained while still keeping up with regular business affairs. However, once the kinks were worked out and employees overcame the learning curve, Grainger rebounded with characteristic strength and fortitude, recording a 7 percent increase in both sales and earnings for 2000.65

Consistent with its values, the company did not lay blame for failures in the system or the transition; rather, its position was to take responsibility for any failures and hold itself accountable. "That took an enormous amount of courage," said George Rimnac. "We had people in the branches who were just tearing themselves up trying to provide customer service the best they could under those terrible circumstances, which is what they always did any time we had a change that was disrupting. Everyone has always pulled together, and it's that kind of high integrity that causes people to want to work for an organization like Grainger."66

Above right: Dick Quast, vice president for real estate, began working at Grainger in 1957. Quast was most instrumental in acquiring land for and overseeing construction of the new headquarters in Lake Forest, Illinois. He retired from the company in 2002.

Right: In 2001, Jim Ryan was promoted from president of Grainger.com to executive vice president for marketing and sales. His career with Grainger began in 1980.

A New Home

In the midst of these developments, Grainger was deeply involved in building its new headquarters in Lake Forest, Illinois, a Chicago suburb. Though the construction itself was relatively painless, the years spent planning the new facility were filled with turmoil.

Recognizing that the company would continue its rapid growth and seeing how quickly Grainger had outgrown its headquarters in Niles, Jere Fluno began talking to David Grainger about buying more land. "I had been working on David for years to buy a tract of land," Fluno said. "Whether we ever built on it or not, we should buy land so that we could put it in our back pocket and then maybe trade it for something else. Finally, after chasing David for many, many years he told me that if I could convince the board, then go ahead."[67]

Fluno did convince the board, and in 1986 the company conducted an employee demographic study to determine the best location for a new headquarters. Finding good real estate that was close enough to the interstate was challenging, but finally Fluno saw what looked like an ideal piece of land in Lake Forest. The company's real estate department had already rejected the land, but Fluno had a keen financial mind and recognized a good investment when he saw it. "We thought we could work with the village of Mettawa in buying all the property, that we could work out an arrangement whereby all of us could win," said Fluno.[68]

After conferring with the village of Mettawa's officials about rezoning for an office complex, Grainger had purchased 510 acres in Mettawa in May 1988 with the intent of developing a plan that would maintain the beauty and character of the area and protect the environment as well. (The purchase of additional land in 1990 brought the total to 525 acres.)

"Before we bought the property, we talked informally to the village mayor, the park preserve, and others to see if they'd be receptive to grant a change in zoning for Grainger to build its office there if we screened it from view," said Jim Baisley, who started up Grainger's legal department in 1981 and retired as senior vice president and general counsel in 2000. "The feedback was positive. We weren't looking for promises, but we tested the waters and the waters seemed fine."[69]

"Normally we don't buy property without the right zoning, but this was an excellent property," added Dick Quast, who, as head of Grainger's real estate department, was instrumental in the purchase. "It was not only beautiful in terms of the amenities on the site itself. It was strategically located to the expressway system and to where our employees live. At that time, we were running about six separate office buildings in the greater Chicago area, and the intent was to consolidate most of these basically into one location."[70]

Unfortunately, after Grainger purchased the property and spent more than a year working with the planning commission, the residents of Mettawa began voicing their objections.

"The land was basically used as farmland and wooded area, so the Mettawa residents mounted a very severe objection to what we wanted to do," said Quast. "However, the property was unique at that time in that it was not totally surrounded geographically by the village of Mettawa. On two sides of the property was Lake County, which was a county zoning jurisdiction, so we had the ability to legally disconnect from Mettawa and put the 500 acres back into the county."[71]

Though the county circuit court disconnected the Grainger land from Mettawa, the village requested a series of appeals through various courts. All appeals were rejected, with the Illinois Supreme Court issuing the final rejection in December 1993. After that, Grainger applied for zoning to the Lake County board and won, despite a last-minute letter-writing campaign from opponents.[72]

Construction finally began in 1996, a full 10 years after Grainger had conducted its demographic study and nine years after it had purchased the property. When completed in 1999, Grainger's beautiful new headquarters sat amid a true conservation development preserving the open space and rural character of the area. Grainger's 650,000-square-foot office campus, comprising a three-story and a four-story building connected by an atrium, was carefully positioned over a 155-acre parcel of land and screened from offsite view. The buildings were designed so that additions could be built higher and still be screened from view.

136   THE LEGEND OF GRAINGER

More importantly, Grainger had proven that it was possible to develop the land while remaining sensitive to the environment. Grainger donated 257 acres known as Grainger Woods Conservation Preserve to the Lake County Forest Preserve. And all of this new preserve was created at no cost to the public. In total, more than 300 of the 525 acres were kept as permanent open space, including the entire frontage along Highway 60.[73]

"Today we have a great relationship with the village of Mettawa," said Jim Baisley. "But this whole adventure meant that, for the first time, we had to deal with the media and with the public. I think we came through with flying colors. Many employees in Lake County wrote to newspaper editors and talked to neighbors about Grainger, explaining what we were doing."[74]

Many of Grainger's people were involved in driving the Lake Forest project to completion, but the project would not have been possible without the skill and perseverance of three key individuals. Jere Fluno, through strategic thinking and long-term planning, had initiated the purchase. Jim Baisley had worked through the seemingly endless legal challenges. And Dick Quast worked on

After years of zoning difficulties, Grainger at last was able to go ahead with the construction of its new headquarters in Lake Forest, Illinois. The finished headquarters featured 650,000 square feet of office space and comprised a three-story and a four-story building.

getting the proper zoning and spent endless hours with architects, landscapers, and construction subcontractors to produce the ideal office campus. "Baisley, Fluno, and Quast carried the mantle over a prolonged period of time and provided special service and dedication to the company to deliver the completed headquarters," said David Grainger.[75]

Once the 1,500 employees had moved into the new headquarters, they understood that it truly had been designed with them in mind. Not only did all employees have ample work space with natural light; they also discovered a

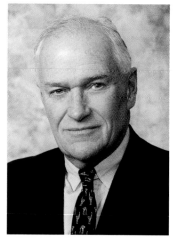

wealth of amenities, including a convenience store, dry-cleaning, photo development, a car service program, a fitness-and-wellness center, an automatic teller machine, a job training center, and a 500-seat auditorium.[76]

As Grainger had intended from the beginning, consolidating people into the new headquarters improved communication throughout the company. "It's a very social building," said Keyser nearly a year after moving to the new headquarters. "It's exceeded my expectations."[77]

Wes Clark pointed out that while the building's design encouraged people to interact, it was not just "social for social's sake. It's more of a give-and-take," he said. "When you have a top-down management structure, the only place the organization comes together is at the top. We're trying to get it so that the top actually works for the people who work here, so that it's a team process that works horizontally at all levels."[78]

The company's founder, immortalized in a bronze relief located just inside the main entrance, would have been truly amazed. The new head-

---

Above: In 2000, James Baisley retired as Grainger's senior vice president and general counsel after 18 years of dedicated service.

Below: Grainger donated 257 acres of undeveloped land to Lake County. The tract became known as Grainger Woods Conservation Preserve.

David Grainger, in his 50th year with the company, and his wife, Juli, celebrate New Year's Eve 2000 in Vienna, Austria.

quarters was a stark contrast to the tiny loft where Grainger started out in September 1927.

Digital Endeavors

Starting in 1999, Grainger began investing in other digital initiatives in addition to Grainger.com, creating such Web sites as Orderzone.com and TotalMRO.com, on-line superstores that brought multiple distributors together under one umbrella; FindMRO.com, which helped customers find hard-to-locate or oddball MRO products or services; and MROverstocks.com, where Grainger posted discontinued or excess inventory at greatly reduced prices. Later, the company merged Orderzone.com with a similar small-business marketplace called Works.com and retained a 40 percent stake in the new entity.

By the end of 1999, Grainger's Internet initiatives had begun receiving plaudits in some of the media's most respected forums. CNN reported that Grainger's presence on the Web solidified its position as one of the largest business-to-business distributors of MRO supplies in the world. In November 1999, *Fortune* magazine featured an article that named Grainger one of "10 Companies That Get It," based on experts' opinions of which big businesses "have e-plays that work." The magazine had nothing but praise for Grainger's foresight and innovation.[79]

Other industry experts, too, were impressed with Grainger's innovative strides into e-commerce. "Grainger has done a great job," said John Sviokla, vice chairman of a Chicago-based digital-strategy consulting firm. "They have used the Net to increase services and sales and market penetration. They have a lot of guts."[80]

Anthony Paoni, a clinical professor of technology and e-commerce, said Grainger showed how well an older company could embrace new technology. "A dinosaur has been taught to dance," he said.[81]

In 1999, Grainger's Internet sales swelled by more than 650 percent over the previous year to

New Tricks for an Old Company

THROUGHOUT ITS HISTORY, GRAINGER has remained on the cutting edge of technology, updating its computer systems and software to ensure maximum efficiency for both customers and employees. Earl Pope managed Grainger's Tabulating department in the 1950s, when records were maintained on punch cards. Ed Bender, who spearheaded many of the company's information technology developments, was another important figure in Grainger's computer-system evolution. As Bender noted years after his retirement in 1994, "I loved all that new stuff. That was the fun I had working there, and I got paid to do it."[1]

In 2002, Earl Pope and Ed Bender took a trip down memory lane to chronicle the company's information technology prior to the ERP system that Grainger converted to in 1999.

1950: The Tabulating department maintains records of each branch's sales and an inventory of each stock number on Kardex cards. Manual sorting of thousands of paper invoices is tedious and time consuming.

1954: Grainger begins keypunching tab cards for each line on an invoice, and the cards are sorted on a sorting machine. A tabulating machine produces hard copy used to post to the Kardex records.

1956: Grainger invests in an interpreter machine that prints on the cards what the holes mean, thus eliminating the time-consuming tabulator step.

1957: The company gets its first Univac, a vacuum-tube machine capable of calculating at a rate equal to three of the department's previous machines.

1961: Grainger upgrades to the Univac Solid State, capable of reading tab cards, calculating, printing reports, and punching tab cards.

1966: Grainger begins the long process of converting to the IBM 360, which uses magnetic tape rather than punch cards.

Late 1960s: Grainger installs a new input device called the IBM 1287, which scans and processes a user's handwriting.

Mid-1970s: The IBM 3741 replaces the 1287 in Grainger's branches. The 3741s are sophisticated keypunch data processors that store information and transmit it to the main computer in Chicago each night. Grainger also installs one of the first sophisticated software inventory management programs in its main headquarters computer. Called Impact, it analyzes sales and inventory levels and determines which items should be ordered and in what quantity.

1976: Grainger replaces the IBM 3741 with programmable Data Point computers in all its branches. This family of computers comes in various sizes and levels of sophistication, thus allowing Grainger to match the appropriate computer to each branch's sales volume and swap the computer for larger or smaller ones without having to rewrite all the programs.

> **Early 1980s:** Grainger begins transitioning from Data Points to the IBM Series 1, which represents the next level in data processing sophistication.
>
> **1983:** Grainger installs its internally developed, computerized, branch order processing and inventory control system (BOP/ICS) in all of its branches.
>
> **1984–1985:** Grainger installs an internally developed computer system called LINQ (local inventory inquiry) in all its branches, improving service and sales and lowering inventory costs.
>
> **1987:** Grainger upgrades the computer system once again with the IBM RS6000. In another step to improve inventory control, warehouse employees begin using handheld terminals that communicate with the local computer system.
>
> **1990:** Grainger speeds up sales transactions by replacing phone lines with a satellite network that links its branches and distribution centers.
>
> **1991:** Grainger introduces its internally developed Electronic Catalog on CD-ROM.
>
> **1995:** Grainger.com debuts and soon allows customers to order via the Internet.
>
> **1996:** In another step to improve inventory management, Grainger begins using a Smart-LINQ system to reduce overstock inventory. The company also develops a system that helps branches stock products for local requirements.
>
> **1999:** Grainger implements a new enterprise resource planning (ERP) system in time for the year 2000 rollover.

$102 million. Some of that business may well have been cannibalized from Grainger's more traditional catalog and branch sales, but a 2000 study showed that Grainger's average order size on traditional channels was $130 but that customers were spending nearly twice that, $240, when they purchased at Grainger.com.[82]

Still, most Grainger customers used the Internet to buy only one-third of their MRO supplies, which meant Grainger enjoyed a competitive advantage: To procure the remaining two-thirds of supplies, customers could order by phone using Grainger's catalogs or walk into any one of its hundreds of branches.[83]

Industry experts widely agreed that it was just that sort of diversity that would separate the winners from the losers. "Will Web clicks chip away at the electrical industry's proud bricks?" asked *Electrical Wholesaling* magazine in April 2000. "The latest thinking is that the winners of these Web wars will be built with a mix of 'clicks-and-mortar' that offer the Web's instant ordering capabilities and the dependable local delivery, on-staff expertise, and warehousing that the best distributors now offer."[84]

If that didn't describe Grainger to a tee, nothing could. *Business Week* printed an interview with Dick Keyser in June 2000 and said the CEO had "one foot in the Old Economy [bricks and mortar] and one in the New [e-commerce]" and that Grainger had "been a pioneer in on-line business-to-business transactions."[85]

The Hills of Wall Street

Unfortunately, as technology stocks, especially dot-coms, suffered a meltdown in 2000, the more traditional segment of Wall Street was becoming impatient for Grainger's Internet initiatives to pay off. Though by the third quarter of 2000 Grainger's on-line sales were up 400 percent from the previous year, the company's overall profits had been pinched by the substantial start-up investments in its digital strategy.[86]

P. Ogden Loux, who was promoted to senior vice president and chief financial officer in 1996, rightly pointed out that, when dealing with e-commerce, "Time is critical. The pace at which e-commerce will change is pretty rapid, and we've had to keep up at an equally fast pace. It would have been nice to invest at a slower pace, but if we did that, and e-commerce continued to move so quickly, we'd be dead."[87]

In January 2001, Grainger consolidated FindMRO.com, MROverstocks.com, and TotalMRO.com into a separate organization called Material Logic, led by Don Bielinski. The new entity sought participation from other industrial distributors and investors, but it seemed that business-to-business Internet adoption would take longer than many in the industry, including Grainger, had expected. In May 2001, Material Logic was discontinued, and Bielinski, after a 30-year career with the company, took early retirement to pursue a career in business-to-business Internet and technology-related areas.[88]

"Though the solution was sound [for Material Logic], the timing for investment and customer adoption was ahead of the market," Keyser explained. "Economic and market conditions made it tough to find funding partners."[89]

The company also sold its 40 percent stake in Works.com, taking a $38 million write-down for its discontinued digital strategies. FindMRO.com was reintegrated with the core, branch-based business segment.[90]

The first three words of Grainger's 2000 annual report were "Back to basics..." Analysts approved of the company's strategy. "You're back to the new old Grainger," said an analyst with Robert W. Baird & Company, of Milwaukee, who, after hearing of Grainger's decision to pull back from some of its Internet strategies, immediately upgraded the company's stock.[91] An analyst with Morningstar Inc., a Chicago provider of financial research, called Grainger's decision "good news" since it would ultimately make the company more profitable.[92]

One Company, Unified Goals

Even after discontinuing some of its digital strategies, Grainger had one foot firmly in the burgeoning realm of e-commerce and another in the traditional bricks-and-mortar business of providing MRO solutions to all types of customers. Though the two methods of doing business were certainly distinct, Keyser made sure that people throughout the company knew that the essential goals of Grainger remained unified.

"Keyser has done a very good job of getting people to focus on the idea that we're all in this together and that you shouldn't think about e-commerce as something separate," said John Howard. "The e-commerce business is just done a different way."[93]

The challenge, as Howard saw it, would be staying ahead of the curve. But as it had done for decades, Grainger was constantly reexamining itself and changing its processes to ensure that it didn't lose its competitive position.

At the April 2001 shareholders' meeting, Keyser reiterated that the Internet would continue to be an important part of the company's growth strategy and explained that Grainger had defined a new meaning for B2B: back to basics. "We're simplifying our operations and reinforcing the quality of service we provide to our customers," he said. "That requires integrating our bricks with our clicks and focusing on how the two work together."[94]

The Industry Standard

One of Grainger's great strengths had always been its willingness to constantly reexamine itself. In the face of the fragmented MRO market, in 2001 Grainger reorganized again to better coordinate its business processes and improve operations. The reorganization also allowed the company to focus on its core growth opportunities so that it would be better able to gain market share. As Keyser told investors in April 2001, "Our compound growth rate over the last 10 years is 9.8 percent. We believe it's possible to continue that trend."[95]

The organizational changes started with Wes Clark adding the role of company president and chief operating officer to his slate of responsibilities. The U.S. marketing, sales, branch network, and e-business groups were integrated under the jurisdiction of James Ryan, who moved from his role as president of Grainger.com to become executive vice president. Grainger's total U.S. sales force was combined into a single nationwide group under Robert Thrush, who was named vice president of sales. Donna Broome was named vice president for national accounts and government sales; Pat Davidson became vice president for branch services; Kathy Hebb became vice president for marketing; and Carl Turza was named vice president for e-business.

In addition, Grainger's quality, business systems, and information resources had been consolidated into a single, company-wide function under the leadership of Tim Ferrarell, who was promoted to senior vice president for enterprise systems. Acklands - Grainger Inc. shifted to the leadership of John Schweig, senior vice president for business development and international. Other executive changes included recruiting Peter Perez to join Grainger as senior vice president of human resources.

In addition to restructuring its operations, Grainger planned to accelerate growth by increasing on-line sales through Grainger.com and by utilizing the Internet to cut costs associated with internal communication and communication with suppliers. The company also planned to continue serving as a critical inventory stocking partner for customers' specific products, which helped it stay strong during economic recessions. Moreover, Grainger would target specific, high-growth customer segments such as the U.S. government.[96]

As the company approached its 75th anniversary, it seemed that all aspects of Grainger's businesses had either completed rebuilding themselves or were in the process of doing so. The company's continued efforts at expense control were helping improve the bottom line.

Even with all the changes the company had experienced, the intent behind its operations remained the same: Grainger helped companies of all sizes to excel through a variety of cost-saving techniques such as inventory management, on-line ordering, product standardization, supplier consolidation, and work-flow integration. By creating innovative processes, embracing new technologies, keeping a strong balance sheet, and honing the skills of its talented employees, Grainger was able to secure its position as the leading North American source of broad-line MRO supplies.

Of course, Grainger's top executives wisely realized that the company could lose its leadership position if they let themselves become passive or complacent. "It's a slippery slope," said Clark. "We have to keep investing and changing so that we can hold and improve our position."[97]

He continued: "As we look forward, there are four things we need to work toward. First, we have to be simple so that it's intuitively obvious to the customer what we do and what we stand for. We also have to become more useful to our customers so when they call up and don't really know what it is they need, we can provide more technical solutions, more wrap-around products, more services than we ever have before. Next, we have to be easy. The business is getting more complex, and somehow we have to hang on to being easy to work with. We have to keep being speedy and convenient. Finally, we have to become more process-oriented to improve our cost structure, and the Web is helping us do that."[98]

Despite such challenges, Dick Keyser remained certain of Grainger's future. He probably summed up the company's confidence and caliber best when he told investors, "We know the MRO business better than anyone else, and we are using that experience to make each layer of our business the standard for the industry.... Whether customers 'click' onto Grainger.com, 'call' a service center representative, or 'stop by' a branch, we believe that Grainger is the only company in the industry that will deliver consistent, multi-channel sales and service to its customers."[99]

APPENDIX A: OFFICERS OF W. W. GRAINGER, INC.

Name	Years Served
William W. Grainger	1927–1968
Margaret L. Grainger	1928–1953
Verna O. Ward	1933–1941
Edward F. Schmidt	1938–1973
Elmer O. Slavik	1938–1948
Russell M. Francis	1953–1970
David W. Grainger	1953–
Howard D. Alder	1967–1975
Walter S. Booth	1967–1972
W. Paul Burch	1967–1975
Gerald F. Donohue	1967–1977
Herbert O. Elfstrom	1967–1969
Lee J. Flory	1967–1991
Robert H. Lollar	1967–1982
Robert W. Wiggins	1967–1983
Edwin C. Zimmer	1967–1977
Ervin W. Palluth	1970–1981
Richard D. Quast	1970–2002
Robert D. Scanlan	1970
Jere D. Fluno	1971–2000
Daniel A. Gregorich	1973–1974
Donald W. Hansen	1973–1981
Richard F. Norman	1973–1985
Max C. Mielecki	1975–1994
Wiley N. Caldwell	1977–1992
George T. Mathews	1978–1985
Robert D. Pappano	1978–2001
Frank J. Juranek	1978–1980
Richard E. Alsch	1978
James P. Dries	1978–1985
Carl E. Frank	1978–1982
James H. Windsor	1978–1988
Robert J. Slobig	1978–1985
David K. Barth	1979–1989
Edward C. Bender	1979–1994
Neal Ormond	1979–1994
Raymond R. Greabe	1979–1994
Dennis G. Guth	1979–1991
Richard H. Hantke	1979–1996
David A. Gall	1979–1985
William J. Jarvis	1980–1985
James M. Baisley	1981–2000
Jack D. Carbone	1982–1985
Donald E. Bielinski	1983–2001
Donald A. Donato	1983–1984
John J. Rozwat	1983–1996
Bernard R. Berntson	1984
Timothy W. Hennessey	1984
Micheal G. Murray	1985–2001
Robert A. Collins	1986–1993
Richard L. Keyser	1986–
P. Ogden Loux	1987–
Paul J. Wallace	1989–2002
John S. Slayton	1989–
Robert J. Gariano	1989–1994
John A. Schweig	1990–
Lester L. Balick	1990–1992
Robert M. Brudzinski	1990–1992
Fred E. Loepp	1990–
Mark S. Hoffman	1991
Jerry D. Wallace	1991
Michael R. Kight	1992–
Peggy H. Stich	1992–1994
Wesley M. Clark	1992–
James T. Ryan	1993–
Barbara M. Chilson	1994–1998
Timothy M. Ferrarell	1994–
Douglas E. Witt	1994–1996
Gary J. Goberville	1995–2001
Rick L. Adams	1996–
Yang C. Chen	1996–
Douglass C. Cumming	1996–1999
Dennis G. Jensen	1996–2001
Larry J. Loizzo	1996–
Philip M. West	1996–
Peter J. Torrenti	1996–2000
Robert A. Thrush	1996–
James M. Tenzillo	1996–2000
Daniel M. Hamburger	1998–1999
Benedetto Randazzo	1998–
Patrick H. Davidson	1999–
Edward J. Franczek	1999–2001
Douglas J. Harrison	1999–
Nancy A. Hobor	1999–
John L. Howard	1999–
Richard J. Reese	1999–2000
George C. Rimnac	1999–
Robert S. Wasserman	1999
Steven J. Borre	2000–2001
Michael G. DeCata	2000–2001
Kathleen R. Hebb	2000–
John C. Milazzo	2000–2001

APPENDIX A: OFFICERS OF W. W. GRAINGER, INC.

Name	Years Served	Name	Years Served
Judith E. Andringa	2002–	Philip A. Lippert	2002–
Mark W. Bischoff	2002–	Bonnie J. McIntyre	2002–
Donna Broome	2002–	Peter M. Perez	2002–
Ronald L. Jadin	2002–	Michael A. Pulick	2002–
Jarnail S. Lail	2002–	James M. Roots	2002–

APPENDIX B: EXECUTIVE COUNCIL OF W. W. GRAINGER, INC., AS OF MAY 2002

DAVID W. GRAINGER
Sr. Chairman of the Board

RICHARD L. KEYSER
Chairman of the Board and
Chief Executive Officer

WESLEY M. CLARK
President and Chief Operating Officer

TIMOTHY M. FERRARELL
Sr. VP, Enterprise Systems

NANCY A. HOBOR
VP, Communications
and Investor Relations

JOHN L. HOWARD
Sr. VP and General Counsel

LARRY J. LOIZZO
VP; President, Lab Safety Supply

P. OGDEN LOUX
Sr. VP, Finance, and
Chief Financial Officer

PETER M. PEREZ
Sr. VP, Human Resources

JAMES T. RYAN
Executive VP,
Marketing, Sales, and Service

JOHN A. SCHWEIG
Sr. VP, Business Development
and International

JOHN W. SLAYTON JR.
Sr. VP, Supply Chain Management

APPENDIX C: OFFICERS OF W. W. GRAINGER, INC., AS OF MAY 2002

RICK L. ADAMS
VP, Supply Chain
Development

JUDITH E. ANDRINGA
VP and Controller

MARK BISCHOFF
VP, Distribution Operations

DONNA BROOME
VP, National Accounts
and Governmental Sales

Y. C. CHEN
VP, Supply Chain Services

PATRICK H. DAVIDSON
VP, Branch Services

DOUGLAS J. HARRISON
President,
Acklands - Grainger Inc.

KATHLEEN R. HEBB
VP, Marketing

RONALD L. JADIN
VP, Finance

MICHAEL R. KIGHT
VP, Grainger Integrated
Supply Operations

JARNAIL S. LAIL
VP, Business Systems

PHILIP A. LIPPERT
VP, Administrative Services

FRED E. LOEPP
VP, Production
Management

BONNIE J. MCINTYRE
VP, Special Product
Services

MICHAEL A. PULICK
VP, Product Management

BENEDETTO RANDAZZO
VP, Latin America

GEORGE C. RIMNAC JR.
VP and Chief Technologist

JAMES M. ROOTS
VP, Marketing Services

ROBERT A. THRUSH
VP, Sales

PHILIP M. WEST
VP and Treasurer

APPENDIX D: DIRECTORS OF W. W. GRAINGER, INC.

Company Incorporated on December 27, 1928

Director — Year elected a Director

Director	28	29	30	31	32	33	34	35	36	37	38	39	40	41	42
William W. Grainger*	•	•	•	•	•	•	•	•	•	•	•	•	•	•	•
Hally Ward Grainger	•	•	•	•	•	•	•	•	•	•	•	•	•	•	•
Margaret L. Grainger	•	•	•	•	•	•									
Verna O. Ward						•	•	•	•	•					
Elmer O. Slavik											•	•	•	•	•
Barbara Grainger Tresemer															
Edward F. Schmidt*															
Josephine Schmidt															
John Wendall Ward															
Antoinette Slavik															
Elmer R. Slavik															
Vernon Tittle															
David W. Grainger															
Graydon H. Ellis															
John E. Jones															

Company Goes Public on March 29, 1967

Director	67	68	69	70	71	72	73	74	75	76	77	78	79	80	81
William W. Grainger*	•	•	•	•	•	•	•	•	•	•	•	•	•	•	•
Elmer O. Slavik				•											
Edward F. Schmidt*	•	•	•	•	•	•	•	•	•	•					
Elmer R. Slavik	•	•	•	•	•	•	•	•	•	•	•	•	•	•	•
David W. Grainger	•	•	•	•	•	•	•	•	•	•	•	•	•	•	•
Graydon H. Ellis	•	•	•	•	•	•	•	•	•	•	•	•	•	•	•
John E. Jones	•	•	•												
Maurice H. Stans	•	•													
Lee A. Doerr			•	•	•	•									
Kingman Douglass		•	•	•	•	•	•	•	•	•	•	•	•	•	•
Thomas L. Doerr					•	•									
Frank W. Norris							•								
Jere D. Fluno									•	•	•	•	•	•	•
Donald W. Hansen									•	•	•	•	•	•	•
George R. Baker										•	•	•	•	•	•
James W. Button													•		
Wiley N. Caldwell													•	•	•
Richard D. Black														•	•
Edward W. Duffy*														•	•
George T. Mathews														•	•
Harold B. Smith, Jr.															•
Robert E. Elberson															
Fred L. Turner															
Wilbur H. Gantz															
James D. Slavik															
John W. McCarter															
Richard L. Keyser															
Janiece S. Webb															
Brian P. Anderson															
Neil S. Novich															
Wesley M. Clark															
Frederick A. Krehbiel															

*Term ended due to director's death.

APPENDIX D: DIRECTORS OF W. W. GRAINGER, INC.

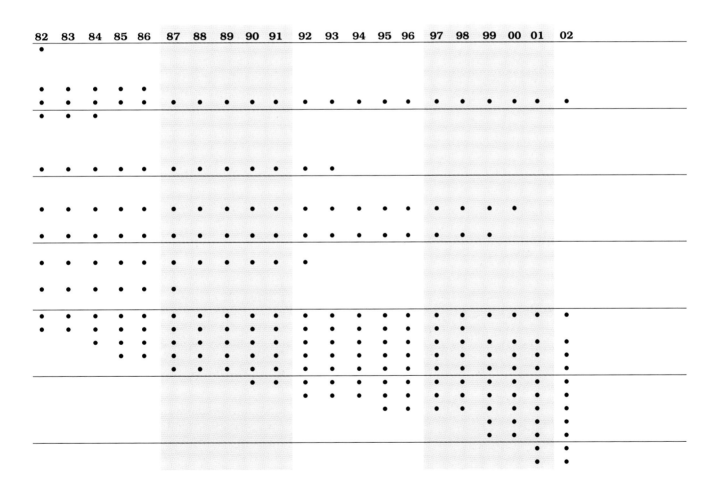

APPENDIX E: STEADY SALES GROWTH THROUGH THE YEARS

SINCE ITS FOUNDING IN 1927, GRAINGER has never experienced an unprofitable year. The $100 million sales level was passed in 1968, and in 1978 it surpassed the half-billion-dollar mark. Grainger crested the billion-dollar milestone in 1984, $2 billion in 1991, $3 billion in 1994, and $4 billion in 1997.

Net Sales of W. W. Grainger, Inc. (000's)

Year	Sales	Year	Sales	Year	Sales	Year	Sales
1928	67	1941	2,647	1954	21,643	1978	564,905
1929	116	1942	3,236	1955	24,516	1979	685,660
1930	161	1943	2,492	1956	25,645	1980	742,949
1931	209	1944	2,926	1957	26,491	1981	822,587
1932	163	1945	3,561	1958	27,661	1982	773,477
1933	236	1946	5,071	1959	33,257	1983	848,462
1934	354	1947	6,128	1960	34,709	1984	1,022,526
1935	558	1948	7,802	1961	38,030	1985	1,091,590
1936	967	1949	8,474	1962	43,533	1986	1,159,595
1937	1,340	1950	12,987	1963	48,774	1987	1,320,797
1938	1,284	1951	16,723	1964	55,277	1988	1,535,468
1939	1,367	1952	18,352	1965	65,059	1989	1,727,454
1940	1,703	1953	21,105	1966	80,226	1990	1,966,815
				1967	88,579	1991	2,111,909
				1968	103,346	1992	2,410,059
				1969	122,880	1993	2,678,981
				1970	132,232	1994	3,083,410
				1971	143,097	1995	3,344,064
				1972	180,404	1996	3,616,640
				1973	227,075	1997	4,226,941
				1974	281,164	1998	4,438,975
				1975	304,370	1999	4,636,275
				1976	371,858	2000	4,977,044
				1977	466,195	2001	4,754,317

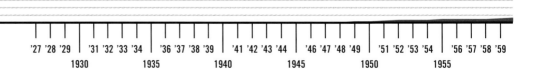

APPENDIX E: STEADY SALES GROWTH THROUGH THE YEARS 151

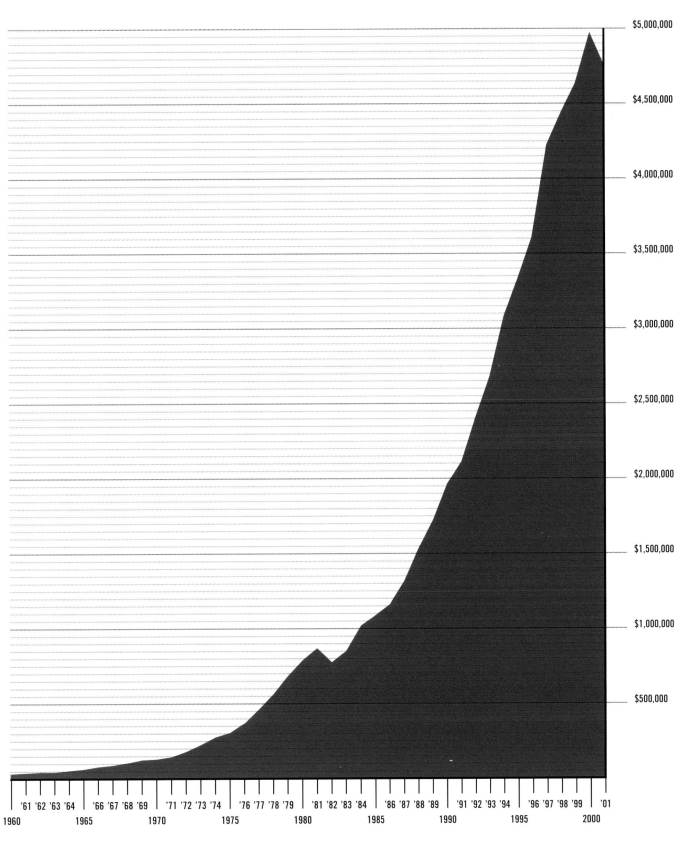

Notes to Sources

Chapter One

1. David Grainger, interview by the author and Richard Buskin, tape recording, 17 May 2000, Write Stuff Enterprises.
2. David Grainger, interview by the author, tape recording, 15 April 1998, Write Stuff Enterprises.
3. Richard Quast, interview by the author, tape recording, 15 April 1998, Write Stuff Enterprises.
4. David Grainger, interview by the author and Richard Buskin, tape recording, 17 May 2000, Write Stuff Enterprises.
5. David Grainger, interview by the author, tape recording, 15 April 1998, Write Stuff Enterprises.
6. David Grainger, interview by the author and Richard Buskin, tape recording, 17 May 2000, Write Stuff Enterprises.
7. David Grainger, interview by the author and Richard Buskin, tape recording, 17 May 2000, Write Stuff Enterprises.

Chapter Two

1. Finis Farr, *Chicago: A Personal History of America's Most American City* (Arlington House), 348–352; Emmett Dedmon, *Fabulous Chicago: A Great City's History and People* (Atheneum), 295–299; Curt Johnson with R. Craig Sautter, *Wicked City - Chicago: From Kenna to Capone* (December Press), 298–300.
2. Finis Farr, *Chicago: A Personal History of America's Most American City* (Arlington House), 370–372.
3. Finis Farr, *Chicago: A Personal History of America's Most American City* (Arlington House), 372–373.

Chapter Three

1. Emmett Dedmon, *Fabulous Chicago: A Great City's History and People* (Atheneum), 300.

2. Finis Farr, *Chicago: A Personal History of America's Most American City* (Arlington House), 373–374.
3. Robert Cromie, *A Short History of Chicago* (Lexicos), 119–120.
4. David Grainger, interview by Max Mielecki, tape recording, 24 October 1996, W. W. Grainger, Inc.
5. David Grainger, interview by the author, tape recording, 15 April 1998, Write Stuff Enterprises.
6. George Brown Tindall and David Emory Shi, *America*, brief fourth edition (New York: W. W. Norton & Company, 1997), 847.
7. David Grainger, interview by the author and Richard Buskin, tape recording, 17 May 2000, Write Stuff Enterprises.
8. David Grainger, interview by the author, tape recording, 15 April 1998, Write Stuff Enterprises.
9. David Grainger, interview by the author and Richard Buskin, tape recording, 17 May 2000, Write Stuff Enterprises.
10. David Grainger, interview by he author and Richard Buskin, tape recording, 17 May 2000, Write Stuff Enterprises.
11. Lee Flory, interview by Anthony Wall, tape recording, 17 May 2000, Write Stuff Enterprises.
12. David Grainger, interview by the author and Richard Buskin, tape recording, 17 May 2000, Write Stuff Enterprises.
13. James Slavik, interview by the author, tape recording, 12 June 2000, Write Stuff Enterprises.
14. James Slavik, interview by the author, tape recording, 12 June 2000, Write Stuff Enterprises.
15. Letter from E. F. Schmidt, 28 December 1934, E. F. Schmidt correspondence file, W. W. Grainger, Inc., archives.
16. Letter from E. F. Schmidt, 28 December 1934, E. F. Schmidt correspondence file, W. W. Grainger, Inc., archives.
17. E. F. Schmidt correspondence file, W. W. Grainger, Inc., archives.
18. E. F. Schmidt correspondence file, W. W. Grainger, Inc., archives.
19. Letter from E. F. Schmidt to Margaret Grainger, 20 February 1935, E. F. Schmidt correspondence file, W. W. Grainger, Inc., archives.
20. David Grainger, interview by the author, tape recording, 15 April 1998, Write Stuff Enterprises.
21. "PVB," interview by Max Mielecki, tape recording, 10 October 1996, W. W. Grainger, Inc.
22. E. F. Schmidt correspondence file, W. W. Grainger, Inc., archives.
23. E. F. Schmidt correspondence file, W. W. Grainger, Inc., archives.
24. E. F. Schmidt correspondence file.
25. E. F. Schmidt correspondence file, W. W. Grainger, Inc., archives.
26. David Grainger, interview by the author, tape recording, 15 April 1998, Write Stuff Enterprises.

27. Memo to W. W. Grainger, Inc., employees, 14 March 1939, W. W. Grainger, Inc., archives.
28. Memo to W. W. Grainger, Inc., employees, 27 March 1939, W. W. Grainger, Inc., archives.
29. Memo to W. W. Grainger, Inc., employees, 27 March 1939, W. W. Grainger, Inc., archives.

Chapter Four

1. Franklin D. Roosevelt, radio address, NBC and CBS radio networks, 5 September 1939.
2. *Chicago Days: 150 Defining Moments in the Life of a Great City* (Chicago: *Chicago Tribune*), 163.
3. *Chicago Days: 150 Defining Moments in the Life of a Great City* (Chicago: *Chicago Tribune*), 163.
4. Finis Farr, *Chicago: A Personal History of America's Most American City* (Arlington House), 392–393.
5. Letter from E. F. Schmidt, 19 October 1940, E. F. Schmidt correspondence file, W. W. Grainger, Inc., archives.
6. David Grainger, interview by the author and Richard Buskin, tape recording, 17 May 2000, Write Stuff Enterprises.
7. David Grainger, interview by the author and Richard Buskin, tape recording, 17 May 2000, Write Stuff Enterprises.
8. David Grainger, interview by the author, tape recording, 15 April 1998, Write Stuff Enterprises.
9. David Grainger, interview by the author, tape recording, 15 April 1998, Write Stuff Enterprises.
10. Nancy Thurber, interview by the author, tape recording, 31 July 2000, Write Stuff Enterprises.
11. Edward C. Bender, interview by the author, tape recording, 24 August 2000, Write Stuff Enterprises.
12. David Grainger, interview by Max Mielecki, tape recording, 24 October 1996, W. W. Grainger, Inc.
13. David Grainger, interview by Max Mielecki, tape recording, 24 October 1996, W. W. Grainger, Inc.
14. Memo to W. W. Grainger, Inc., employees, 15 February 1943, W. W. Grainger, Inc. archives.
15. Letter from E. F. Schmidt to E. O. Slavik, 1943, W. W. Grainger, Inc., archives.

Chapter Five

1. Finis Farr, *Chicago: A Personal History of America's Most American City* (Arlington House), 394.
2. Emmett Dedmon, *Fabulous Chicago: A Great City's History and People* (Atheneum), 348.
3. Memo from E. F. Schmidt, 29 September 1945, E. F. Schmidt correspondence file, W. W. Grainger, Inc., archives.
4. Memo from E. F. Schmidt, 29 September 1945, E. F. Schmidt correspondence file, W. W. Grainger, Inc., archives.

5. Memo from E. F. Schmidt, 29 September 1945, E. F. Schmidt correspondence file, W. W. Grainger, Inc., archives.
6. E. O. Slavik, "Resume of Grainger Policy," October 1945, W. W. Grainger, Inc., archives.
7. Robert Wiggins, interview by Max Mielecki, tape recording, 14 October 1996, W. W. Grainger, Inc.
8. Information provided by Philip Lippert to Melody Maysonet, 26 March 2002.
9. David Grainger, interview by the author and Richard Buskin, tape recording, 17 May 2000, Write Stuff Enterprises.
10. James Slavik, interview by the author, tape recording, 12 June 2000, Write Stuff Enterprises.
11. "Guide to Purchasing," 18 January 1949, W. W. Grainger, Inc.
12. David Grainger, interview by the author, tape recording, 17 November 1998, Write Stuff Enterprises.
13. David Grainger, interview by the author, tape recording, 17 November 1998, Write Stuff Enterprises.

Chapter Six

1. Melvyn Dubofsky, Athan Theoharis, and Daniel M. Smith, *The United States in the Twentieth Century* (Prentice-Hall, 1978); Hugh Brogan, *The Longman History of the United States of America* (Longman Group, 1985); David Manning White, *Popular Culture: Mirror of American Life* (Publisher's Inc., 1977); J. Ronald Oakley, *God's Country: America in the Fifties* (Dembner Books, 1986); Carl Solberg, *Riding High—America in the Cold War* (Mason & Lipscomb, 1973).
2. David Grainger, interview by the author and Richard Buskin, tape recording, 17 May 2000, Write Stuff Enterprises.

Chapter Seven

1. Melvyn Dubofsky, Athan Theoharis, and Daniel M. Smith, *The United States in the Twentieth Century* (Prentice-Hall, 1978); Hugh Brogan, *The Longman History of the United States of America* (Longman Group, 1985); David Farber, *The Age of Great Dreams: America in the 1960s* (HarperCollins, 1994).
2. Dale Woods, interview by Max Mielecki, tape recording, n.d., W. W. Grainger, Inc.
3. Rich Greenlee, interview by Max Mielecki, tape recording, 10 March 1997, W. W. Grainger, Inc.
4. Angie Salazar, interview by the author, tape recording, 24 August 2000, Write Stuff Enterprises.
5. Richard Quast, interview by the author, tape recording, 15 April 1998, Write Stuff Enterprises.
6. Max Mielecki, interview by the author, tape recording, 24 August 2000, Write Stuff Enterprises.
7. Michael Kight, interview by the author, tape recording, 17 November 1998, Write Stuff Enterprises.

8. Dale Woods, interview by Max Mielecki, tape recording, n.d., W. W. Grainger, Inc.
9. David Grainger, interview by the author, tape recording, 17 November 1998, Write Stuff Enterprises.
10. Richard Quast, interview by the author, tape recording, 15 April 1998, Write Stuff Enterprises.
11. Edward C. Bender, interview by Max Mielecki, tape recording, 23 October 1996, W. W. Grainger, Inc.
12. Edward C. Bender, interview by Max Mielecki, tape recording, 23 October 1996, W. W. Grainger, Inc.
13. David Grainger, interview by the author and Richard Buskin, tape recording, 17 May 2000, Write Stuff Enterprises.
14. E. F. Schmidt correspondence file, W. W. Grainger, Inc., archives.
15. Max Mielecki, interview by the author, tape recording, 24 August 2000, Write Stuff Enterprises.
16. David Grainger, interview by the author and Richard Buskin, tape recording, 17 May 2000, Write Stuff Enterprises.
17. David Grainger, interview by the author and Richard Buskin, tape recording, 17 May 2000, Write Stuff Enterprises.
18. David Grainger, interview by the author and Richard Buskin, tape recording, 17 May 2000, Write Stuff Enterprises.
19. David Grainger, interview by the author and Richard Buskin, tape recording, 17 May 2000, Write Stuff Enterprises.
20. James Slavik, interview by the author, tape recording, 12 June 2000, Write Stuff Enterprises.
21. David Grainger, interview by the author and Richard Buskin, tape recording, 17 May 2000, Write Stuff Enterprises.
22. Jere Fluno, interview by the author, tape recording, 15 April 1998, Write Stuff Enterprises.
23. Lee Flory, interview by Anthony Wall, tape recording, 17 May 2000, Write Stuff Enterprises.
24. W. W. Grainger, Inc., 1967 Annual Report, 1.
25. W. W. Grainger, Inc., 1968 Annual Report, 2.
26. Edwin Darby, *Chicago Sun-Times*, 12 July 1968.
27. W. W. Grainger, Inc., 1968 Annual Report, 4.
28. W. W. Grainger, Inc., 1969 Annual Report, 3.
29. E. F. Schmidt, interview by Newstrack, Inc., 30 May 1970, W. W. Grainger, Inc., archives.

Chapter Eight

1. George B. Tindall and David E. Shi, *America*, brief second edition (New York: W. W. Norton & Company, 1989), 911–912.
2. "Inflation and Deflation," *Microsoft Encarta 97 Encyclopedia.*
3. Eric Hobsbawm, *Age of Extremes—The Short Twentieth Century* (Michael Joseph, 1994); Melvyn Dubofsky, Athan Theoharis, and Daniel M. Smith, *The United States in the*

Twentieth Century (Prentice-Hall, 1978); Hugh Brogan, *The Longman History of the United States of America* (Longman Group, 1985).
4. W. W. Grainger, Inc., 1971 Annual Report, 2.
5. W. W. Grainger, Inc., 1973 Annual Report, 5.
6. W. W. Grainger, Inc., 1971 Annual Report, 2–3.
7. W. W. Grainger, Inc., 1975 Annual Report, 2, 3.
8. W. W. Grainger, Inc., 1976 Annual Report, 10.
9. Robert Collins, interview by Max Mielecki, tape recording, November 1996, W. W. Grainger, Inc.
10. *W. W. Grainger, Inc., 50 Years of Growth, 1927–1977*, 15.
11. George Rimnac, interview by the author, tape recording, 31 July 2000, Write Stuff Enterprises.
12. James Slavik, interview by the author, tape recording, 12 June 2000, Write Stuff Enterprises.
13. Jere Fluno, interview by the author, tape recording, 15 April 1998, Write Stuff Enterprises.
14. W. W. Grainger, Inc., 1977 Annual Report, 2–3.
15. *W. W. Grainger, Inc., 50 Years of Growth, 1927–1977*, 4.
16. W. W. Grainger, Inc., 1978 Annual Report, 6–7.
17. W. W. Grainger, Inc., 1979 Annual Report, 8.

Chapter Nine

1. John Taylor, *Circus of Ambition: The Culture of Wealth and Power in the Eighties* (Warner Books, 1989); Vincent Virga, *The Eighties: Images of America* (HarperCollins, 1992).
2. W. W. Grainger, Inc., 1980 Annual Report, ii.
3. W. W. Grainger, Inc., 1985 Annual Report, iv.
4. W. W. Grainger, Inc., 1982 Annual Report, ii.
5. William Baldwin, "Dollars from Doodads," *Forbes*, 11 October 1982.
6. William Baldwin, "Dollars from Doodads," *Forbes*, 11 October 1982.
7. W. W. Grainger, Inc., 1984 Annual Report, v.
8. Max Mielecki, interview by the author, tape recording, 24 August 2000, Write Stuff Enterprises.
9. W. W. Grainger, Inc., 1984 Annual Report, v.
10. George Rimnac, interview by the author, tape recording, 31 July 2000, Write Stuff Enterprises.
11. John Slayton, interview by the author, tape recording, 11 September 1997, Write Stuff Enterprises.
12. W. W. Grainger, Inc., 1982 Annual Report, ii.
13. Wiley Caldwell, interview by the author, tape recording, 12 June 2002, Write Stuff Enterprises.
14. Wiley Caldwell, interview by the author, tape recording, 12 June 2002, Write Stuff Enterprises.
15. *Corporate Focus*, December 1982.
16. *Network*, January 1983.
17. James Baisley, interview by the author, tape recording, 11 September 1997, Write Stuff Enterprises.
18. *Network*, January 1983.
19. Wiley Caldwell, interview by the author, tape recording, 12 June 2002, Write Stuff Enterprises.

20. W. W. Grainger, Inc., 1984 Annual Report, v.
21. *Network,* January/February 1985.
22. W. W. Grainger, Inc., 1985 Annual Report, v.
23. H. Lee Murphy, "Grainger Boosts Distribution System," *Crain's Chicago Business,* 5 May 1986.
24. Wiley Caldwell, interview by the author, tape recording, 12 June 2002, Write Stuff Enterprises.
25. W. W. Grainger, Inc., 1981 Annual Report, iii.
26. W. W. Grainger, Inc., 1983 Annual Report, v.
27. Andrew Thomas, interview by Anthony Wall, tape recording, 17 May 2000, Write Stuff Enterprises.
28. Wiley Caldwell, interview by the author, tape recording, 12 June 2002, Write Stuff Enterprises.
29. Joanne Cleaver, "Reinvigorated Grainger Beefs Up Distribution," *Crain's Chicago Business,* 4 May 1987.
30. H. Lee Murphy, "Grainger Boosts Distribution System," *Crain's Chicago Business,* 5 May 1986.
31. Nancy Thurber, interview by the author, tape recording, 31 July 2000, Write Stuff Enterprises.
32. H. Lee Murphy, "Grainger Boosts Distribution System," *Crain's Chicago Business,* 5 May 1986.
33. Joanne Cleaver, "Reinvigorated Grainger Beefs Up Distribution," *Crain's Chicago Business,* 4 May 1987.
34. W. W. Grainger, Inc., 1987 Annual Report, v.
35. Joanne Cleaver, "Reinvigorated Grainger Beefs Up Distribution," *Crain's Chicago Business,* 4 May 1987.
36. W. W. Grainger, Inc., 1987 Annual Report, iv, xi.
37. Wiley Caldwell, interview by the author, tape recording, 12 June 2002, Write Stuff Enterprises.
38. James Slavik, interview by the author, tape recording, 12 June 2000, Write Stuff Enterprises.
39. Matt O'Connor, "Surprise of a Market Has Grainger Growing," *Chicago Tribune,* 12 July 1987.
40. John Schweig, interview by the author, tape recording, 24 May 1999, Write Stuff Enterprises.
41. Andrew Thomas, interview by Anthony Wall, tape recording, 17 May 2000, Write Stuff Enterprises.
42. Peter D. Waldstein, "Aggressive Expansion Gives Grainger New Growth Thrust," *Crain's Chicago Business,* 5 October 1987.
43. Robert Thrush, interview by the author, tape recording, 11 September 1997, Write Stuff Enterprises.
44. Matt O'Connor, "Surprise of a Market Has Grainger Growing," *Chicago Tribune,* 12 July 1987.
45. *Network,* January/February 1987.
46. Joanne Cleaver, "Reinvigorated Grainger Beefs Up Distribution," *Crain's Chicago Business,* 4 May 1987.
47. Peter D. Waldstein, "Aggressive Expansion Gives Grainger New Growth Thrust," *Crain's Chicago Business,* 5 October 1987.

48. Matt O'Connor, "Surprise of a Market Has Grainger Growing," *Chicago Tribune*, 12 July 1987.
49. Matt O'Connor, "Surprise of a Market Has Grainger Growing," *Chicago Tribune*, 12 July 1987.
50. Steven R. Strahler, "Grainger Earns High Marks for Distribution Strategy," *Crain's Chicago Business*, 15 February 1988.
51. Joanne Cleaver, "Reinvigorated Grainger Beefs Up Distribution," *Crain's Chicago Business*, 4 May 1987.
52. Peter D. Waldstein, "Aggressive Expansion Gives Grainger New Growth Thrust," *Crain's Chicago Business*, 5 October 1987.
53. W. W. Grainger, Inc., 1989 Annual Report, iv.
54. Joanne Cleaver, "'Plain-Jane' Grainger Sitting Pretty, Still in Growth Mode," *Crain's Chicago Business*, 22 May 1989.
55. Charles Siler, "The Goal is 0 Percent," *Forbes*, 30 October 1989.
56. W. W. Grainger, Inc., 1987 Annual Report, viii.
57. Matt O'Connor, "Surprise of a Market Has Grainger Growing," *Chicago Tribune*, 12 July 1987.
58. Charles Siler, "The Goal is 0 Percent," *Forbes*, 30 October 1989.
59. W. W. Grainger, Inc., 1988 Annual Report, ii.
60. W. W. Grainger, Inc., 1987 Annual Report, xi.
61. Joanne Cleaver, "'Plain-Jane' Grainger Sitting Pretty, Still in Growth Mode," *Crain's Chicago Business*, 22 May 1989.
62. Joanne Cleaver, "'Plain-Jane' Grainger Sitting Pretty, Still in Growth Mode," *Crain's Chicago Business*, 22 May 1989.
63. W. W. Grainger, Inc., 1989 Annual Report, ii.
64. "Area Service Companies Fill Fortune Ranks," *Crain's Chicago Business*, 12 June 1989.

Chapter Ten

1. Jere Fluno, interview by the author, tape recording, 17 May 2000, Write Stuff Enterprises.
2. Micheal Murray, interview by the author, tape recording, 1 May 1997, Write Stuff Enterprises.
3. Wesley Clark, interview by Anthony Wall, tape recording, 17 May 2000, Write Stuff Enterprises.
4. W. W. Grainger, Inc., 1995 Annual Report, 2.
5. W. W. Grainger, Inc., 1991 Annual Report, vi.
6. John Schweig, interview by the author, tape recording, 24 May 1999, Write Stuff Enterprises.
7. Donald Bielinski, interview by the author, tape recording, 17 November 1998, Write Stuff Enterprises.
8. W. W. Grainger, Inc., 1988 Annual Report, ii.
9. W. W. Grainger, Inc., 1992 Annual Report, xv.
10. W. W. Grainger, Inc., 1993 Annual Report, 13.
11. W. W. Grainger, Inc., 1990 Annual Report, xiii.
12. Peter Torrenti, interview by the author, tape recording, 15 April 1998, Write Stuff Enterprises.
13. W. W. Grainger, Inc., 1992 Annual Report, xiii.
14. W. W. Grainger, Inc., 1992 Annual Report, xvii.
15. Donald Bielinski, interview by the author,

tape recording, 17 November 1998, Write Stuff Enterprises.
16. Wesley Clark, interview by the author, tape recording, 24 May 1999, Write Stuff Enterprises.
17. Fred Loepp, interview by the author, tape recording, 17 November 1998, Write Stuff Enterprises.
18. W. W. Grainger, Inc., 1992 Annual Report, xi.
19. Larry Loizzo, interview by the author, tape recording, 24 August 2000, Write Stuff Enterprises.
20. W. W. Grainger, Inc., 1992 Annual Report, xix.
21. W. W. Grainger, Inc., 1995 Annual Report, 15.
22. James Tenzillo, interview by the author, tape recording, 16 November 1998, Write Stuff Enterprises.
23. W. W. Grainger, Inc., 1992 Annual Report, xxi.
24. W. W. Grainger, Inc., 1994 Annual Report, 17.
25. W. W. Grainger, Inc., 1995 Annual Report, 17.
26. Nancy Thurber, interview by the author, tape recording, 31 July 2000, Write Stuff Enterprises.
27. Mike Dorsher, "Behind the Money: A Man Who Acts on Opportunities," *Wisconsin State Journal*, 29 August 1993.
28. Jere Fluno, interview by the author, tape recording, 17 May 2000, Write Stuff Enterprises.
29. Jere Fluno, interview by the author, tape recording, 15 April 1998, Write Stuff Enterprises.
30. Benedetto Randazzo, interview by the author, tape recording, 24 May 1999, Write Stuff Enterprises.
31. Richard Keyser, interview by the author, tape recording, 1 May 1997, Write Stuff Enterprises.
32. Richard Keyser, interview by the author, tape recording, 16 May 2000, Write Stuff Enterprises.
33. W. W. Grainger, Inc., 1990 Annual Report, vi.
34. Edward C. Bender, interview by Max Mielecki, tape recording, 23 October 1996, W. W. Grainger, Inc.
35. W. W. Grainger, Inc., 1990 Annual Report, iv.
36. W. W. Grainger, Inc., 1992 Annual Report, ix.
37. W. W. Grainger, Inc., 1994 Annual Report, 11.
38. Andy Cohen, "Practice Makes Profits: Sales Training Spurs Double-Digit Increases Every Year for W. W. Grainger," *Sales & Marketing Management*, July 1995.
39. Jack Keough, "W. W. Grainger: Focusing on the Customer Ahead of the Competition," *Supply Chain Yearbook 2000*.
40. Robert Pappano, interview by the author, tape recording, 16 November 1998, Write Stuff Enterprises.
41. John Slayton, interview by the author, tape recording, 11 September 1997, Write Stuff Enterprises.
42. James Baisley, interview by the author, tape recording, 11 September 1997, Write Stuff Enterprises.
43. Thomas Malak and Joseph Malak, interview by Melody Maysonet, tape recording, 11 April 2002, Write Stuff Enterprises.
44. Thomas Malak and Joseph Malak, interview by Melody

Maysonet, tape recording, 11 April 2002, Write Stuff Enterprises.
45. Donald Bielinski, interview by the author, tape recording, 17 November 1998, Write Stuff Enterprises.
46. Jere Fluno, interview by the author, tape recording, 15 April 1998, Write Stuff Enterprises.
47. Timothy Ferrarell, interview by the author, tape recording, 16 November 1998, Write Stuff Enterprises.
48. Fred Loepp, interview by the author, tape recording, 17 November 1998, Write Stuff Enterprises.
49. Michael Tellor, interview by Richard F. Hubbard, tape recording, 10 May 2001, Write Stuff Enterprises.
50. Tracy MacMillan, interview by Richard F. Hubbard, tape recording, 10 May 2001, Write Stuff Enterprises.
51. Jim Lindemann, interview by Richard F. Hubbard, tape recording, 14 May 2001, Write Stuff Enterprises.
52. Fritz Zeck, interview by Richard F. Hubbard, tape recording, 26 April 2001, Write Stuff Enterprises.
53. Mark MacDonald, interview by Richard F. Hubbard, tape recording, 26 April 2001, Write Stuff Enterprises.
54. W. W. Grainger, Inc., 1995 Annual Report, 5.
55. John Slayton, interview by the author, tape recording, 11 September 1997, Write Stuff Enterprises.
56. James Tenzillo, interview by the author, tape recording, 16 November 1998, Write Stuff Enterprises.
57. W. W. Grainger, Inc., 1990 Annual Report, ix.
58. W. W. Grainger, Inc., 1994 Annual Report, 11.
59. Richard Keyser, interview by the author, tape recording, 1 May 1997, Write Stuff Enterprises.
60. W. W. Grainger, Inc., 1994 Annual Report, 7.
61. David Young, "W. W. Grainger Agrees on a Three-Way Alliance," *Chicago Tribune*, 5 January 1996.
62. Mike Dorsher, "Behind the Money: A Man Who Acts on Opportunities," *Wisconsin State Journal*, 29 August 1993.

Chapter Eleven

1. W. W. Grainger, Inc., 1999 Annual Report, 5.
2. Barbara Chilson, interview by the author, tape recording, 15 April 2000, Write Stuff Enterprises.
3. "Survey Finds Corporate Purchasers Plan to Increase Use of Internet," PR Newswire, 13 November 1997.
4. George Rimnac, interview by the author, tape recording, 31 July 2000, Write Stuff Enterprises.
5. "Grainger Internet Commerce Customers Can Now Check Local Product Availability and Customer-Specific Pricing Online," PR Newswire, 17 September 1997.
6. "When Natural Disaster Strikes grainger.com Provides On-line Emergency Information," PR Newswire, 16 July 1998.
7. Sara Procknow, "How Does Your Web Site

Rate?" *Industrial Distribution*, 31 August 1998.
8. "Grainger Recognized as Leader in Information Technology," PR Newswire, 24 August 1998.
9. Patrick Davidson, interview by Melody Maysonet, tape recording, 19 September 2000, Write Stuff Enterprises.
10. David Grainger, interview by the author, tape recording, 17 November 1998, Write Stuff Enterprises.
11. Rick Adams, interview by the author, tape recording, 16 November 1998, Write Stuff Enterprises.
12. W. W. Grainger, Inc., 1996 Annual Report, 9, 10, 13.
13. W. W. Grainger, Inc., 1997 Annual Report, 7.
14. Grainger 2000 Fact Book, 11.
15. Wesley Clark, interview by the author, tape recording, 24 May 1999, Write Stuff Enterprises.
16. W. W. Grainger, Inc., 2000 Annual Report, 7.
17. Wesley Clark, interview by the author, tape recording, 24 May 1999, Write Stuff Enterprises.
18. Wesley Clark, interview by the author, tape recording, 24 May 1999, Write Stuff Enterprises.
19. "Grainger Chairman and CEO Reiterates Guidance for 2001 and Provides 2002 Outlook to Analysts," PR Newswire, 30 November 2001.
20. "Wesley M. Clark Addresses Analysts at Baird 2000 Industrial Technology Conference on Grainger's Internet Strategy," Business Wire, 15 November 2000.
21. Larry Loizzo, interview by the author, tape recording, 24 August 2000, Write Stuff Enterprises.
22. W. W. Grainger, Inc., 1996 Annual Report, 18.
23. W. W. Grainger, Inc., 1998 Annual Report, 17.
24. W. W. Grainger, Inc., 1996 Annual Report, 18.
25. W. W. Grainger, Inc., 1998 Annual Report, 15.
26. "Grainger Subsidiary Lab Safety Supply Acquires Direct Marketing Company," PR Newswire, 26 February 2001.
27. Donald Bielinski, interview by the author, tape recording, 17 November 1998, Write Stuff Enterprises.
28. Peter Torrenti, interview by the author, tape recording, 15 April 1998, Write Stuff Enterprises.
29. "Large Companies Identify Three Ways to Improve Indirect Materials Management," PR Newswire, 10 December 1998.
30. Wesley Clark, interview by Anthony Wall, tape recording, 17 May 2000, Write Stuff Enterprises.
31. Michael Kight, interview by the author, tape recording, 17 November 1998, Write Stuff Enterprises.
32. Wesley Clark, interview by Anthony Wall, tape recording, 17 May 2000, Write Stuff Enterprises.
33. W. W. Grainger, Inc., 1997 Annual Report, 15.
34. Benedetto Randazzo, interview by the author, tape recording, 24 May 1999, Write Stuff Enterprises.
35. Benedetto Randazzo, interview by the author, tape recording, 24 May

1999, Write Stuff Enterprises.
36. Wesley Clark, interview by Anthony Wall, tape recording, 17 May 2000, Write Stuff Enterprises.
37. *Newslink*, January 1998, Acklands - Grainger, Inc.
38. Douglas Cumming, interview by the author, tape recording, 23 August 2000, Write Stuff Enterprises.
39. Richard Keyser, interview by the author, tape recording, 1 May 1997, Write Stuff Enterprises.
40. "W. W. Grainger, Inc. Reaches Definitive Agreement to Acquire the Canadian Industrial Distribution of Acklands Limited," PR Newswire, 6 November 1996.
41. Richard Keyser, interview by the author, tape recording, 1 May 1997, Write Stuff Enterprises.
42. James Baisley, interview by the author, tape recording, 11 September 1997, Write Stuff Enterprises.
43. W. W. Grainger, Inc., 1997 Annual Report, 9, 10.
44. "A Strategy Update... Defining the Key Elements of Our Plan," *Newslink*, September 2000.
45. Douglas Harrison, interview by the author, tape recording, 23 August 2000, Write Stuff Enterprises.
46. Y. C. Chen, interview by the author, tape recording, 31 July 2000, Write Stuff Enterprises.
47. John Schweig, interview by the author, tape recording, 24 May 1999, Write Stuff Enterprises.
48. Y. C. Chen, interview by the author, tape recording, 31 July 2000, Write Stuff Enterprises.
49. W. W. Grainger, Inc., 1996 Annual Report, 3.
50. Gary Goberville, interview by the author, tape recording, 24 May 1999, Write Stuff Enterprises.
51. Paul Wallace, interview by the author, tape recording, 16 November 1998, Write Stuff Enterprises.
52. John Howard, interview by the author, tape recording, 17 May 2000, Write Stuff Enterprises.
53. Jere Fluno, interview by the author, tape recording, 15 April 1998, Write Stuff Enterprises.
54. Nancy Hobor, interview by the author, tape recording, 31 July 2000, Write Stuff Enterprises.
55. Dick Heiman, interview by Richard F. Hubbard, tape recording, 26 April 2001, Write Stuff Enterprises.
56. Joe Ramos, interview by Richard F. Hubbard, tape recording, 10 May 2001, Write Stuff Enterprises.
57. Nancy Hobor, interview by the author, tape recording, 31 July 2000, Write Stuff Enterprises.
58. Nancy Thurber, interview by the author, tape recording, 31 July 2000, Write Stuff Enterprises.
59. Richard Quast, interview by Anthony Wall, tape recording, 17 May 2000, Write Stuff Enterprises.
60. "W. W. Grainger, Inc., Recognized by *Fortune* as One of 'The 100 Best Companies to Work for in America,'"

PR Newswire, 19 December 1997.
61. "W. W. Grainger, Inc., Recognized by *Fortune* as One of 'The 100 Best Companies to Work for in America,'" PR Newswire, 19 December 1997.
62. James Ryan, interview by the author, tape recording, 16 November 1998, Write Stuff Enterprises.
63. W. W. Grainger, Inc., 1999 Annual Report, 6.
64. Patrick Davidson, interview by Melody Maysonet, tape recording, 19 September 2000, Write Stuff Enterprises.
65. "Grainger Reports EPS of 51 Cents for the 2000 Fourth Quarter and $2.05 for the Full Year," Business Wire, 30 January 2001.
66. George Rimnac, interview by the author, tape recording, 31 July 2000, Write Stuff Enterprises.
67. Jere Fluno, interview by the author, tape recording, 17 May 2000, Write Stuff Enterprises.
68. Jere Fluno, interview by the author, tape recording, 17 May 2000, Write Stuff Enterprises.
69. James Baisley, interview by the author, tape recording, 24 May 1999, Write Stuff Enterprises.
70. Richard Quast, interview by the author, tape recording, 15 April 1998, Write Stuff Enterprises.
71. Richard Quast, interview by the author, tape recording, 15 April 1998, Write Stuff Enterprises.
72. Ted Kleine, "Open Space, Developers Can Coexist," *Chicago Tribune*, 11 June 1996.
73. James Baisley, "The Other Side of Grainger's Plan," *Chicago Tribune*, 4 September 1994.
74. James Baisley, interview by the author, tape recording, 24 May 1999, Write Stuff Enterprises.
75. Information provided by Philip Lippert to Melody Maysonet, 26 March 2002.
76. Mark Spencer, "New W. W. Grainger Building Tough to Find, but Not for Long," *Chicago Daily Herald*, 3 June 1999.
77. Richard Keyser, interview by the author, tape recording, 16 May 2000, Write Stuff Enterprises.
78. Wesley Clark, interview by Anthony Wall, tape recording, 17 May 2000, Write Stuff Enterprises.
79. Melanie Warner, "10 Companies That Get It," *Fortune*, 8 November 1999.
80. Bob Black, "Booming B2B: 'Brick-to-Click'; Grainger's Huge Catalog Goes on Web," *Chicago Sun-Times*, 7 February 2000.
81. Bob Black, "Booming B2B: 'Brick-to-Click'; Grainger's Huge Catalog Goes on Web," *Chicago Sun-Times*, 7 February 2000.
82. Grainger 2000 Fact Book, 7.
83. W. W. Grainger, Inc., 1999 Annual Report, 12.
84. Jim Lucy, "Why Bricks Will Still Matter," *Electrical Wholesaling*, April 2000.
85. "Junk That Catalog and Get on the Web," *Business Week*, 26 June 2000.
86. Vance Cariaga, "W. W. Grainger's Web Investments: Money

Well Spent? Don't Ask Street," *Investor's Business Daily*, 19 September 2000.
87. P. Ogden Loux, interview by the author, tape recording, 17 May 2000, Write Stuff Enterprises.
88. "Grainger Reorganizes to Focus on Core Growth Opportunities," PR Newswire, 1 June 2001.
89. Richard L. Keyser, comments at W. W. Grainger Annual Meeting of Shareholders, 25 April 2001.
90. James P. Miller, "Grainger Closing Money-Losing Web Subsidiary," *Chicago Tribune*, 24 April 2001.
91. Julie Johnson, "Tech Watch: An E-consulting Guru Marches Back in Time," *Crain's Chicago Business*, 30 April 2001.
92. Howard Wolinsky, "Grainger Cuts Web Sites," *Chicago Sun-Times*, 24 April 2001.
93. John Howard, interview by the author, 17 May 2000, Write Stuff Enterprises.
94. Richard Keyser, comments at W. W. Grainger Annual Meeting of Shareholders, 25 April 2001.
95. Richard Keyser, comments at W. W. Grainger Annual Meeting of Shareholders, 25 April 2001.
96. Richard Keyser, comments at W. W. Grainger Annual Meeting of Shareholders, 25 April 2001.
97. Wesley Clark, interview by Anthony Wall, tape recording, 17 May 2000, Write Stuff Enterprises.
98. Wesley Clark, interview by Anthony Wall, tape recording, 17 May 2000, Write Stuff Enterprises.
99. Richard Keyser, comments at W. W. Grainger Annual Meeting of Shareholders, 26 April 2000; Richard Keyser, comments at W. W. Grainger Annual Meeting of Shareholders, 25 April 2001.

Chapter Eleven Sidebar: The Grainger Name

1. Bob Gideon, interview by the author, tape recording, 17 May 2000, Write Stuff Enterprises.
2. Dennis Jensen, interview by the author, tape recording, 15 April 1998, Write Stuff Enterprises.
3. Judann Pollack, "Grainger Debuts on TV to Tout Industrial 'Stuff,'" *Advertising Age*, 10 November 1997.
4. Judann Pollack, "Grainger Debuts on TV to Tout Industrial 'Stuff,'" *Advertising Age*, 10 November 1997.

Chapter Eleven Sidebar: New Tricks for an Old Company

1. Edward C. Bender, interview by Melody Maysonet, tape recording, 15 April 2002, Write Stuff Enterprises.

INDEX

Page numbers in italics indicate photographs.

A

Acklands, Dudley, 125
Acklands, J. D., 125
Acklands - Grainger Inc., 124–129, *125,* 125, 143
Acklands Ltd., 124–129
acquisitions
    Allied Safety Supply Co., 101, 103
    Ball Industries, 100–101
    Ben Meadows Co., 121
    Bossert Co., 100
    C. L. Gransden & Co., 100
    Doerr Electric Corp., 69
    Jones Safety Supply, 101
    Lab Safety Supply, 103
    Mansco-Lakeshore, 100
    McMillan Manufacturing Co., 73
    Rice Safety Equipment, 101
    Satterlee Co., 100
    SMS Supply Co., 100
    Vonnegut Industrial Products, 100
    W. M. Pattison Supply Co., 100
Adams, Rick, 118
advertising, *82,* 95, *99,* 111, *126*
Advertising Age's Business Marketing, 116
agriculture products, 33, *33,* 38, *38,* 54
Alder, Howard, 53
Alexander Grant & Co., 13, 21, 65–67
alliance partnerships, 113
Allied Safety Supply Co., 101, 103
American Airlines, 98
American Chili Co., 34
awards and recognition
    Advertising Age's Business Marketing, 116
    American Material Management Society Plant of the Year Award, 83
    catalogs, awards for, 103, 116
    Fortune magazine, 95, 97, 133, 139
    H. R. Chally Group World Class Sales Excellence Award, 111
    ISO 9002, 104
    suppliers, awards from, 111

B

Bahnmaier, Emil, 82
Bain and Co., 91
Baisley, James, 84, 86, 97, 108, 135–138, *138*
Baker, George, *93*
Baldor Electric, 19
Barth, David, 91

Bee-Vac clothes washer, 16, *16*
Bender, Edward, 35, 62, 140
Ben Meadows Co., 121
Bielinski, Donald, *90,* 91, 99, 109, 117, 121, 142
Biffle, Greg, *127*
Bissell, Cush, *34*
Blunt Ellis & Loewi, 93
board of directors, 66–67, *92*
Booth, Walt, 53
Bossert Company, 100
Bossert Industrial Supply, 99–100
Brady Corp., 110–111
branches, *80, 110, 119, 123, 143*
   Alaska, 95
   Albany, N.Y., 51
   Atlanta, 22–23
   Baltimore, 51
   Buffalo, N.Y., 35–36
   Canada, 124–129
   Chicago, 52, *61,* 93
   Cincinnati, 35, 51
   Cleveland, 24, *24,* 35
   customized branches, 46–47
   Dallas, 22, 52
   Daytona Beach, Fla., 95
   Denver, 27
   Detroit, 46–47
   expansion and restructuring, 61–62, 93–95, 119–121
   Havana, 43
   Honolulu, 95
   Kansas City, 24, 35, 51, 52
   Los Angeles, 51, 52, 93
   Memphis, 42
   Mexico, 43, 123–124
   Miami, 35–36, 123–124
   Minneapolis, 24
   Morton Grove, Ill., 124
   New Orleans, 27, 35, 51
   New York, 24, 52, *55, 58,* 93
   North Milwaukee, Wis., 69
   Oakland, Calif., 124
   Oklahoma City, 42
   Philadelphia, 20, *21*
   Pittsburgh, 35, 51
   Portland, 42
   Providence, R.I., 35
   Puerto Rico, 106
   Rio de Janeiro, 43
   San Francisco, 22
   Syracuse, N.Y., 56
   Toledo, 42, 51
Broome, Donna, 142
Brown-Brochmeyer, 12
Burke, Wally, *45*
business principles, 16, *18,* 85–86, 130–133
business strategy, 42, 91–95, 98, 143. *See also* customer service; marketing strategy
Business Week, 141
Bussman, 113

C

Cadillac Plastic and Chemical Co., 113
Caldwell, Wiley, 42, *84, 87, 92, 94, 96*
   executive vice president, 84
   order processing system, 83–84
   president of Distribution Group, 78
   succession plan, 91
Campbell Hausfeld, 131–132
catalogs, *117, 118. See also* electronic commerce; *MotorBook*
   Acklands - Grainger Inc., 127
   CD-ROM catalog, 105, *106*
   Dealer Bargain catalogs, *48*
   General Catalog (Grainger), *81,* 105–106, *106*
   Grainger.com, 115–117, *116,* 141
   Grainger's Wholesale Net Price Catalog, 81, 89
   Lab Safety catalogs, *97,* 103, 121, *121*
   Safety Equipment catalog, *90*
   Spanish-language catalog, *124*

Central Distribution Center, 25, *25*, 68, *72–73*, *83*
   automated warehousing, 83
   expansion, 69, 72, 75, 78
   renamed National Distribution Center, 108
Charitable Matching Gift program, 132
Chen, Y. C., 129
Chicago (city), 15, 17, 19–20, 31, 41
Chicago Sun-Times, 69
Chicago Tribune, 31
Chilson, Barbara, 115
CIO, 116
Clark, Wesley, 98, 101, 117, *119*, 119–120, 126
   Acklands Ltd. acquisition, 125–129
   business strategy, 143
   Grainger Custom Solutions, 121
   Lake Forest, Ill., headquarters, 138
   president and chief operating officer, 142
   September 11, 2001, terrorist attacks, 133
Clesco Manufacturing, *13*, 108–109
Cleveland Sales, 108–109
C. L. Gransden & Company, 100
Coca-Cola, 98
College of Customer Contact, 107–108
Collins, Robert, 74
computers. *See* technology
Cooper Lighting Co., 110
Coordinated Sales, 63–64
corporate citizenship, 102, 132–133, 136
Corporate Focus, 85
Crescent Electric Supply Co., 113
Cumming, Douglas, 125
Cummins Engine Co., 91
customer service, 16, *23*, 23–24, 38, 53–54, *85*, 116

D

data processing systems, 73–74, *74*
Davidson, Patrick, 134, 142
Dayton brand motors, 19, 27, *28*, 54, *54*, 73
Dayton Electric Manufacturing Co., 27
Dayton Industries, 90
Delta Air Lines, 98
department head meetings, 36–37
disaster relief, 102, 132–133
distribution centers. *See* Central Distribution Center; regional distribution centers; zone distribution centers
Doerr, Lee, 69
Doerr Electric Corp., 69, 90
Doerr Way, The, 85
Douglass, Kingman, 67, 92
Duffy, Edward, *92*

E

Educational Assistance Ltd., 102
Elberson, Robert, 92
Electrical Wholesaling, 141
electronic commerce, 115–117, *116*, 129, 139–142
Elfstrom, Herb, 53, 67
Ellis, Graydon, 66–67, *92*
Emerson Motor Co., 90, 110
employee benefits, 75, 86–87, 105
   boat trip, 44
   bonuses, 33, 37
   life insurance, 33, 39
   medical insurance, 39, 57, 60
   picnic, 52
   profit sharing, 35, 49, 52, 57, 86–87, 105
   Quota-Makers Holiday, 47, *56*, 57, *71*, 76–77
   scholarships, 60
   vacation time, 57
employee practices, 129–133
   effects of World War II, 32–33

employee practices
    (continued)
    team incentive
        program, 60
    training, 89, 107–108,
        122, 129
employees, 32, 50, 56,
    62, 74, 76–77, 87,
    122
employee stock
    ownership plan,
    86–87
expansion, 25, 61–62,
    72–73, 81. See also
    branches; office
    buildings and land
    branch expansion,
        1987, 93–95
    distribution centers,
        69, 75, 78, 119–120
    international
        distribution, 123–129

F

fans, 15, 42
    agricultural use, 54
    consumer demand,
        28–29, 42–43
    Dayton trademark,
        25
    1920s, use during, 16–17
    parts, 24
    World War II supplies,
        36, 38
Ferguson, Ray, 53
Ferguson Enterprises, 113
Ferrarell, Timothy, 109, 143
Ferrario Co., 51
fiftieth anniversary, 70,
    75, 78

financial data. See sales
    volume
Flory, Lee, 22, 67, 79, 97
Fluno, Jere, 42, 79, 84,
    92, 114, 134
    CD-ROM catalog,
        105–106
    Grainger Inc. values,
        130–131
    Lake Forest head-
        quarters, 135–138
    on legacy of early
        management, 75
    office of the chairman,
        97–98
    public offering of stock,
        66–67
    senior vice president,
        84
    on supplier
        relationships, 109
Forbes magazine, 82, 95
Ford Motor Co., 121
Fortune magazine, 95, 97,
    133, 139
Francis, Rud, 52, 53, 67
Frank Lynn & Associates,
    121

G

General Catalog (Grainger),
    81, 89, 105–106,
    106, 143. See also
    MotorBook
General Electric, 12, 17,
    25–26, 90, 113
General Pump Co., 60
Gideon, Bob, 126
Glore Forgan Wm. R.
    Staats, Inc., 65–67

Golden Dolphin, 63–64, 64
Graham, Mark, 14
Grainger, Barbara, 13, 44
Grainger, David, 42, 45,
    52, 67, 78, 84, 92,
    96, 114, 128, 139
    on Billie's Chili, 34–35
    birth, 13
    business principles,
        85–86
    chairman of the board,
        98, 117
    childhood, 20
    Golden Dolphin, 63–64
    Grainger values, 132
    Grainger, William W.,
        memories of,
        11–13, 19
    joins Grainger full-time,
        52
    office of the chairman,
        97–98
    president, 75, 79
    public offering of stock,
        65–67
    on Quota-Makers
        Holiday, 47
    Schmidt, Edward F.,
        memories of, 21–22
    Slavik, E. O.,
        memories of, 44
Grainger, Hally, 12, 12–13,
    16, 34
Grainger, Juli, 139
Grainger, Laura, 53
Grainger, Margaret, 12,
    13, 20, 24, 53, 53
Grainger, William W.,
    10, 14, 19, 22,
    34, 43, 45, 70,
    78, 115, 128

Grainger, William W., *(continued)*
   business philosophy, 16, *18*
   death, 84–85
   employee relations, 59–60
   founding of W. W. Grainger, Inc., 11–13
   marriage, 12
   personality, 19, 11
   public offering of stock, 65–67
   retirement, 67
   and Schmidt, Edward F., 65
   youth, 11
Grainger Caribe, Inc., 123–124
Grainger Charitable Trust, 60
Grainger.com, 115–117, *116*, 141
Grainger Consulting Services, 108, 117, 121
Grainger Custom Solutions, 117
Grainger Division, 98
Grainger Foundation, Inc., 85
Grainger Global Sourcing, 118
Grainger Grapevine, 40, 44, *49*, 52, 57, 62, 85.
Grainger Industrial Supply, 117, 118–119
Grainger Integrated Supply Operations, 117, 121, *122*
Grainger Parts, 104, 117

Grainger Racing Team, 127, *127*
Grainger, S.A. de C.V., 123–124
Grainger Sanitary Supplies & Equipment, 100–101, *101*
"Grainger's Song," 51
Grainger's Wholesale Net Price Catalog. See General Catalog (Grainger)
Grainger Woods Conservation Preserve, 136, *138*
Great Depression, 17, 19–20
Greenlee, Rich, 59
Green Lights Program, 102

H

Hansen, Don, 79
Harrison, Douglas, 129
headquarters buildings. *See* office buildings and land
Hebb, Kathy, 142
Hedberg, Don, 103
Hedberg, Gerry, 103
Heiman, Dick, 132
Hobor, Nancy, 131–132
Hoch, Frank, *56*
human resources. *See* employee benefits; employee practices

I, J

i.c. stars, 102
Industrial Distribution, 116

international distribution, 43, 123–129
ISO 9002, 104
Jaffe, Mike, 132
Jani-Serv, 99–101
Jensen, Dennis, 126
Johns-Manville, 17
Jones, John E., 66–67
Jones Safety Supply, 101

K

Kasten, Walter, *92*
Kennametal, 113
Keyser, Richard, *114, 128, 134*
   1945 Grainger policy statement, 42
   Acklands Ltd. acquisition, 125
   advertising, 111
   chairman of the board, 117
   e-commerce, 142
   future of W. W. Grainger, Inc., 143
   Grainger values, 132–133
   international expansion, 106
   joins Grainger, 91
   office of the chairman, 97–98
   president, Grainger Division, 97
   September 11, 2001, terrorist attacks, 133
Kight, Mike, 60, 98, 121, *123*

INDEX

Koehler, Fred, 108–109
Korean War, 49–50, 52

L

Labsafety.com, 121
Lab Safety General Catalog, 103
Lab Safety Supply, 103, 117, 121
Lake Forest, Ill., headquarters, *114, 115,* 135–138, *136–137*
Lambert, Jack, 51
Lincoln National Life Insurance Co., 50–51
Lindemann, Jim, 110
Loepp, Fred, 101, 109
logos, *31*
Loizzo, Larry, 103, 121, *121*
Lollar, Robert, 67
Loux, P. Ogden, 91, 141–142

M

MacDonald, Mark, 110–111
MacMillan, Tracy, 109–110
Malak, Joseph, 109
Malak, Thomas, 109
Mansco-Lakeshore, 100
manufacturing operations, 25, 27, 54, 69, 72–73, 89–90
marketing strategy, 87–89, 111, 126–127. *See also* advertising

Marvin, B., *131*
Master motors, 12
Matching Charitable Gifts program, 102
Material Logic, 142
Mathews, George, 78
McCartney, Dave, 132
McDonald's, 98
McKerns, Patricia, 133
McMillan Manufacturing Co., 73
Merrill Lynch, 87
Mexico, expansion into, 123–124
Midwest Stock Exchange, 71
Mielecki, Max, 60, 65, 79, 82
Montgomery Ward, 17
Morningstar Inc., 142
Motion Industries, 113
MotorBook, 17, 18, 43, 46. *See also General Catalog* (Grainger)
  attempt to personalize, 26
  introduction, *12,* 13, 15
  name change, 89
  product offerings, 75
  revised by David Grainger, 52
  1980s, 82–83
motors, *17, 48,* 69
  American motors, 16
  B-Line motors, 16
  Dayton brand, 19, 27, *28,* 54, *54,* 73
  Demco brand, 73
  Dem-Kote brand, 73
  Doerr brand, 73
  Grainger brand, 19

Liberty brand, 51
Speedaire brand, *73*
Sunlight motors, 12, 16
Teel brand, 73
World War II supplies, 36
Murray, Micheal, *98*

N

NASCAR racing, 127
National Association for the Exchange of Industrial Resources, 102
National Distribution Center. *See* Central Distribution Center
National Minority Supplier Development Council Advanced Management Education program, 102
Network, 85
newsletters. *See Grainger Grapevine; Network*
New York Stock Exchange, *66,* 71
Nicholson, Jack, 65–67
NL Industries, 91
Norman, Richard, 79
North Safety, 109–110
Norton Co., 113

O

office buildings and land
  Lake Forest, Ill., *114, 115,* 135–138, *136–137*

office buildings and land *(continued)*
   Loomis Street and Van Buren, 50–51
   Niles, Ill., 61–62, 78
   Oakley/Adams building, *50,* 51
   Skokie, Ill., 78, *88–89*
   South Western Avenue, 21
   22nd Street, 21
   West Adams Street, 50
   West Congress Street, 25, *25*
open accounts, *23,* 23–24

P

Painting Our Future seminar, *130–131*
Paoni, Anthony, 139
Pappano, Robert, 91, 108
Parts Company of America, 99, 101, 103–104, *104*
Peerless, 12
Perez, Peter, 143
Pope, Earl, 140
Power Winch, 131
President's Club, 87, *87*
pricing, 43, 45
products, 15, 69, 75, 82. *See also* fans; motors
   agriculture products, 33, *33,* 38, *38,* 54
   beer pump, *20*
   Bee-Vac clothes washer, 16, *16*
   Billie's Chili, 33–35
   blackout items, 36
   early motors, 15
   gas masks, *39*
   General Electric, 25–26
   Golden Dolphin, 63–64
   Infra-Red Lamp Brooder, 38, *38*
   Korean War, effects of, 49–50
   LumaPro lighting, 118
   milking machine, *33*
   movie cameras, *46*
   Profast'ners brand, 127
   pumps, 60
   refrigeration supplies, 25, 27
   Robur Health Massage Machine, 17
   Shields of Faith Bible, 36, *36*
   Trav-ler Television, *41*
   Westward tools, 118, 127
   wheel arbor adapter, *13*
   wiring devices, 25, 27
   World War II, effects of, 33–38, 42–43
profit sharing, 35, 49, 52, 57, 86–87, 105
public offering of stock, *59,* 65–67, *66*
Puerto Rico branch, 123–124

Q, R

Quast, Dick, 60, 132–133, *134,* 135–138
Quota-Makers Holiday, 47, 56, 57, *71, 75, 76–77*
Randazzo, Ben, 106, 124
regional distribution centers, 120
   Cranford, N.J., 62
   Ft. Worth, Tex., 69
   Greenville, S.C., 95
   Kansas City, 86
   Los Angeles, *120*
   New York, 52, *55,* 56, 62
   Oakland, Calif., 69
Reinier, Bob, *45*
Reinier, Teena, *45*
restructuring, branches, 61–62, 93–95, 119–121
restructuring, corporate, 78–79, 98–104, 117, 142–143
Rice Safety Equipment, 101
Rimnac, George, 74, 83, 116, 134
Ritchey, Royal, 33–34
Robert W. Baird & Company, 93, 142
Robur Health Massage Machine, 17
Royal Chili, 33
Rozwat, John, 84
Rubbermaid, 132
Rust-Oleum, 109
Ryan, James, 133–134, *134,* 142

S

Safety TechLine, 121
Salazar, Angie, 59
Sales & Marketing Management, 108

sales volume
    1930–1939, 20, 27–28
    1948, 44
    1960–1969, 61–62, 67
    1977, 75
    1980–1989, 81, 87–88, 90–91, 95
    1990–1999, 97, 115
Satterlee Co., 100
Schmidt, Edward F., *21, 32, 34, 67*
    Billie's Chili shareholder, 34
    on branches, 23, 24
    on competitive strategy, 26–27
    customer service policies, 38
    executive vice president, 52
    and Grainger, William W., 65
    joins Grainger, 21–22
    payroll policy, 41–42
    public offering of stock, 65–67
    retirement, 75
scholarship program, 60
Schweig, John, 91, 99, 111, 129, 143
Scott Fetzer Co., 90
September 11, 2001, terrorist attacks, 102, 132–133
Shields of Faith Bible, 36, *36*
signs, *11, 58*
Slavik, Antoinette, 34
Slavik, E. O., *22, 45*
    Billie's Chili shareholder, 34
    business strategy, 1945, 42
    death, 74
    department-head meetings, 36–37
    joins Grainger, 22
    retirement, 44–45
Slavik, E. R., 45, 65–67, 91, *93*
Slavik, James, 22, 44–45, 65–67, 74–75, 91
Slavik Printing, 22
Slayton, John, 83, 108, *108*
Smart-LINQ, 118
Smith, Harold Jr., *93*
SMS Supply Co., 100
Specialty Distribution Group, 99–103
Specialty Products Division, 90
Speedaire brand, 73, *73*
Square D, 113
Stans, Maurice, 21, 34, *34,* 65–67
stock splits, 104–105
Streeter, Jack, 44
supplier relationships, 45–46, 64–65, 108–111, 131–132
Sviokla, John, 139
Svitak, Joe, 16

T

Tabulating department, 51, 62, *62*
team incentive program, 60
technology, 94–95, 140–141
    automated order and inventory system, 83–84
    automated warehousing, 78, 83
    computers, early use of, 62–63, *63*
    data processing systems, 73–74
    enterprise resource planning, 133–134
    Kardex records, 51
    satellite network, 107
    WATS service, 62
    Web sites, 115–117, *116,* 121, 129, 139–142
telephone etiquette, 53
Tellor, Michael, 109
Tenzillo, James, 103, 111
Thomas, Andy, 90
Thrush, Robert, 142
Thurber, Nancy, 35, 133
Torrenti, Peter, 100, 121
training, 89, 107–108, *122,* 129
Training and Communications Center, 89
truck fleet, *112*
Turza, Carl, 142

U, V

United Parcel Service, 98
University of Illinois Champaign-Urbana, endowed chair, 84–85

values of W. W. Grainger, Inc., 129–133
Van Bussum, Phil, 53
Velkaborsky, Anne, 16
Vonnegut Industrial Products, 100

W

wages, 41–42
Wagner Electric, 12
Wallace, Paul, 91, 130
Wall Street Journal, 59
Wayne Water Systems, 131
Web sites, 115–117, *116*, 121, 129, 139–141
Westinghouse, 45
Wiggins, Bob, 42–43
Woods, Dale, 59–60
work hours, 27–28, 32–33, 37–39
world events, 31–32. *See also* September 11, 2001, terrorist attacks
World War II, *30*
  effects on Grainger, 32–39, 41–43

X, Y, Z

Zeck, Fritz, 110
Zimmer, Edwin, 67
zone distribution centers, 106, *107*, 119–120